ADVANCED PROBLEMS IN MATHEMATICS

Advanced Problems in Mathematics: Preparing for University

Stephen Siklos

http://www.openbookpublishers.com

© 2016 Stephen Siklos.

This work is licensed under a Creative Commons Attribution 4.0 International license (CC BY 4.0). This license allows you to share, copy, distribute and transmit the work; to adapt the work and to make commercial use of the work providing attribution is made to the author (but not in any way that suggests that they endorse you or your use of the work). Attribution should include the following information:

Stephen Siklos, *Advanced Problems in Mathematics: Preparing for University*. Cambridge, UK: Open Book Publishers, 2016. http://dx.doi.org/ 10.11647/OBP.0075

Further details about CC BY licenses are available at http://creativecommons.org/licenses/by/4.0/

Digital material and resources associated with this volume are available at
http://www.openbookpublishers.com/isbn/9781783741427

STEP questions reproduced by kind permission of Cambridge Assessment Group Archives.

This is the first volume of the OBP Series in Mathematics:

ISSN 2397-1126 (Print)
ISSN 2397-1134 (Online)

ISBN Paperback: 9781783741427
ISBN Digital (PDF): 9781783741441
ISBN Digital ebook (epub): 9781783741458
ISBN Digital ebook (mobi): 9781783741465
DOI: 10.11647/OBP.0075

Cover image: *Paternoster Vents* (2012). Photograph © Diane Potter. Creative Commons Attribution-NonCommercial-NoDerivs CC BY-NC-ND

All paper used by Open Book Publishers is SFI (Sustainable Forestry Initiative) and PEFC (Programme for the Endorsement of Forest Certification Schemes) Certified.

Printed in the United Kingdom and United States by
Lightning Source for Open Book Publishers

Contents

About this book ix

STEP 1

Worked Problems 11
    Worked problem 1   11
    Worked problem 2   15

Problems 19
    P1   An integer equation   19
    P2   Partitions of 10 and 20   21
    P3   Mathematical deduction   23
    P4   Divisibility   25
    P5   The modulus function   27
    P6   The regular Reuleaux heptagon   29
    P7   Chain of equations   31
    P8   Trig. equations   33
    P9   Integration by substitution   35
    P10   True or false   37
    P11   Egyptian fractions   39
    P12   Maximising with constraints   41
    P13   Binomial expansion   43
    P14   Sketching subsets of the plane   45
    P15   More sketching subsets of the plane   47
    P16   Non-linear simultaneous equations   49
    P17   Inequalities   51
    P18   Inequalities from cubics   53
    P19   Logarithms   55
    P20   Cosmological models   57
    P21   Melting snowballs   59
    P22   Gregory's series   61
    P23   Intersection of ellipses   63
    P24   Sketching $x^m(1-x)^n$ 65

P25	Inequalities by area estimates	67
P26	Simultaneous integral equations	69
P27	Relation between coefficients of quartic for real roots	71
P28	Fermat numbers	73
P29	Telescoping series	75
P30	Integer solutions of cubics	77
P31	The harmonic series	79
P32	Integration by substitution	81
P33	More curve sketching	83
P34	Trig sum	85
P35	Roots of a cubic equation	87
P36	Root counting	89
P37	Irrationality of e	91
P38	Discontinuous integrands	93
P39	A difficult integral	95
P40	Estimating the value of an integral	97
P41	Integrating the modulus function	99
P42	Geometry	101
P43	The t substitution	103
P44	A differential-difference equation	105
P45	Lagrange's identity	107
P46	Bernoulli polynomials	109
P47	Vector geometry	111
P48	Solving a quartic	113
P49	Areas and volumes	115
P50	More curve sketching	117
P51	Spherical loaf	119
P52	Snowploughing	121
P53	Tortoise and hare	123
P54	How did the chicken cross the road?	125
P55	Hank's gold mine	127
P56	A chocolate orange	129
P57	Lorry on bend	131
P58	Fielding	133
P59	Equilibrium of rod of non-uniform density	135
P60	Newton's cradle	137
P61	Kinematics of rotating target	139
P62	Particle on wedge	141
P63	Sphere on step	143

P64	Elastic band on cylinder	145
P65	A knock-out tournament	147
P66	Harry the calculating horse	149
P67	PIN guessing	151
P68	Breaking plates	153
P69	Lottery	155
P70	Bodies in the fridge	157
P71	Choosing keys	159
P72	Commuting by train	161
P73	Collecting voles	163
P74	Breaking a stick	165
P75	Random quadratics	167

Syllabus   **169**

About this book

This book has two aims.

- The general aim is to help bridge the gap between school and university mathematics.

  You might wonder why such a gap exists. The reason is that mathematics is taught at school for various purposes: to improve numeracy; to hone problem-solving skills; as a service for students going on to study subjects that require some mathematical skills (economics, biology, engineering, chemistry — the list is long); and, finally, to provide a foundation for the small number of students who will continue to a specialist mathematics degree. It is a very rare school that can achieve all this, and almost inevitably the course is least successful for its smallest constituency, the real mathematicians.

- The more specific aim is to help you to prepare for STEP or other examinations required for university entrance in mathematics. To find out more about STEP, read the next section.

It used to be said that mathematics and cricket were not spectator sports; and this is still true of mathematics. To progress as a mathematician, you have to strengthen your mathematical muscles. It is not enough just to read books or attend lectures. You have to work on problems yourself.

One way of achieving the first of the aims set out above is to work on the second, and that is how this book is structured. It consists almost entirely of problems for you to work on.

The problems are all based on STEP questions. I chose the questions either because they are 'nice' — in the sense that you should get a lot of pleasure from tackling them (I did), or because I felt I had something interesting to say about them.

The first two problems (the 'worked problems') are in a stream of consciousness format. They are intended to give you an idea how a trained mathematician would think when tackling them. This approach is much too long-winded to sustain for the remainder of the book, but it should help you to see what sort of questions you should be asking yourself as you work on the later problems.

Each subsequent problem occupies two pages. On the first page is the STEP question, followed by a comment. The comments may contain hints, they may direct your attention to key points, and they may include more general discussions. On the next page is a solution; you have to turn over, so that your eye cannot accidentally fall on a key line of working. The solutions give enough working for you to be able to read them through and pick up at least the gist of the method; they may not give all the details of the calculations. For each problem, the given solution is of course just one way of producing the required result: there may be many other equally good or better ways. Finally, if there is space on the page after the solution (which is sometimes not the case, especially if diagrams have to be fitted in), there is a postmortem. The postmortems may indicate what aspects of the solution you should be reviewing and they may tell you about the ideas behind the problems.

I hope that you will use the comments and solutions as springboards rather than feather beds. You will only really benefit from this book if you have a good go at each problem before looking at the comment and certainly before looking at the solution. The problems are chosen so that there is something for you to learn from each one, and this will be lost to you for ever if you simply read the solution without thinking about the problem on your own.

I have given each problem a difficulty rating ranging from ✓ to ✓✓✓. Difficulty in mathematics is in the eye of the beholder: you might find a question difficult simply because you overlooked some key step, which on another day you would not have hesitated over. You should not therefore be discouraged if you are stuck on a ✓-question; though you should probably be encouraged if you get through one of the rare ✓✓✓-questions without mishap.

This book is about depth not breadth. I have not tried to teach you any new topics. Instead, I want to lead you towards a deeper understanding of the material you already know. I therefore restricted myself to problems requiring knowledge of the specific and rather limited syllabus that is laid out at the end of the book. The pure mathematics section corresponds to the syllabus for STEP papers I and II. If you are studying British A-levels, Scottish Highers, or the International Baccalaureate, for example, you will be familiar with most of this material. For the mechanics and statistics/probability sections, there is less agreement about what a core syllabus should be (in the IB there is no mechanics at all), so I gave myself a freer hand.

Calculators are not required for any of the problems in this book and calculators are not permitted in STEP examinations. In the early days of STEP, calculators were permitted but they were not required for any question. It was found that candidates who tried to use calculators sometimes ended up missing the point of the question or getting a silly answer. My advice is to remove the battery so that you are not tempted.

I started this section by listing the aims of the book. You may have noticed that teaching you mathematics is not an aim. I can't remember where I heard the following rather nice analogy. In 1464, a huge block of Carrara marble was carefully chosen from a quarry in Tuscany and transported to Florence, where it lay almost untouched for many years. In 1501 it was given to the sculptor Michelangelo. He worked hard on it, chipping away and chipping away for three years, until at last, inside the block, he found a beautiful statue of David. You can see a picture at:

http://www.accademia.org/explore-museum/artworks/michelangelos-david/

And the analogy? I can't teach you mathematics with this book, but I believe that much hard work on your part, chipping away at the problems, will eventually reveal the mathematician that is within you.

I hope you enjoy using this book as much as I have enjoyed putting it together.

STEP

What is STEP?

STEP (Sixth Term Examination Paper) is an examination used by Cambridge University as part of its procedure for admitting students to study mathematics. Applicants are interviewed in December, and may then be offered a place conditional on the results of their public examinations (A-level, International Baccalaureate, etc) and STEP. The examinations are sat in June and offers are confirmed in August when all the examination results are available.

STEP is used for conditional offers not just by Cambridge, but (at the time of writing) also by Warwick University for almost all of its Mathematics offers, and to a lesser extent by some other English universities. Many other university mathematics departments recommend that their applicants practise on the past papers even if they do not take the examination. In 2015, 4322 scripts were marked, only about 1000 of which were written by students holding an offer from Cambridge.

The first STEPs were taken in 1987, and there was were specimen papers before that from which some of the questions in this book were drawn. At that time, there were STEPs in many subjects but by 2001 only the mathematics papers remained. The examination has been more or less stable over nearly 30 years: it has not been blown about by the various fads in the public examinations systems that came and went during that time.

There are three STEPs, called papers I, II and III. Each paper has thirteen questions, including three on mechanics and two on probability/statistics. Candidates are assessed on six questions only.

The pure mathematics question in Papers I and II are based the core A-level Mathematics syllabus, with some minor additions, which is listed at the end of this book. The pure mathematics questions in Paper III are based on a 'typical' Further Mathematics mathematics A-level syllabus (at the time of writing, there is not even a partial core for Further Mathematics A-levels).

There is also no core (at the time of writing) for A-level mechanics and statistics, so the STEP syllabuses for these areas consist of material that a student with a particular interest might have covered. It has to be said, though, that the statistics questions are very likely to require knowledge of probability rather than statistics (for example, there are very few questions on statistical tests of given data). This is because the underlying theory of statistics is quite difficult, and therefore unsuitable for examining at this level, whereas the application of statistical tests is rather routine and again unsuitable for examination at this level.

What is the purpose of STEP?

From the point of view of admissions to a university mathematics course, STEP has three purposes.

- It is used as a hurdle for entrance to university mathematics courses, and sometimes for other mathematics-based courses. There is strong evidence that success in STEP correlates very well

2 *Advanced Problems in Mathematics*

with university examination results.[1]

- It acts as preparation for the university course, because the style of mathematics found in STEP questions is similar to that of undergraduate mathematics.

- It tests motivation. It is important to prepare for STEP (by working through old papers, for example), which can require considerable dedication. Those who are not willing to make the effort are unlikely to thrive on a difficult university mathematics course.

STEP *vs* A-level

A-level[2] tests mathematical knowledge and technique by asking you to tackle fairly stereotyped problems. STEP asks you to apply the same knowledge and technique to problems that are, ideally, unfamiliar.

Here is an A-level question, in which you follow the instructions in the question:

By using the substitution $u = 2x - 1$, or otherwise, find
$$\int \frac{2x}{(2x-1)^2} \, dx.$$

And here, for comparison, is a STEP question, which requires both competence in basic mathematical techniques and mathematical intuition. Note that help is given for the first integral, so that everyone starts at the same level. Then, for the second integral, candidates have to show that they understand why the substitution used in the first part worked, and how it can be adapted.

Use the substitution $x = 2 - \cos\theta$ to evaluate the integral
$$\int_{3/2}^{2} \left(\frac{x-1}{3-x}\right)^{\frac{1}{2}} dx.$$

Show that, for $a < b$,
$$\int_{p}^{q} \left(\frac{x-a}{b-x}\right)^{\frac{1}{2}} dx = \frac{(b-a)(\pi + 3\sqrt{3} - 6)}{12},$$
where $p = (3a+b)/4$ and $q = (a+b)/2$.

The differences between STEP and A-level are:

1. STEP questions are much longer. Candidates completing four questions in three hours will almost certainly get a grade 1.

2. STEP questions are much less routine.

3. STEP questions may require considerable dexterity in performing mathematical manipulations.

4. Individual STEP questions may require knowledge of several different areas of mathematics (especially the mechanics and statistics questions, which will often require advanced pure mathematical techniques).

---

[1] Recent studies comparing rank in STEP with rank in first-year Cambridge mathematics examinations reveal a Spearman correlation coefficient of 0.63, which is very high in comparison with other predictors of university examination results.

[2] I use the term 'A-level' here as a shorthand for a typical school mathematics examination. The particular examinations you take may well be very different in style and format but, even if that is the case, I am sure some of what follows will strike a chord with you.

5. The marks available for each part of the question are not disclosed on the paper.

6. There is a lot of choice on STEP papers (6 questions out of 13).

7. Calculators are not permitted in STEP examinations.

These difference matter, because in mathematics more than in any other subject it is very important to match the difficulty of the question with the ability of the candidates. For example, you could reasonably have the question 'Was Henry VIII a good king?' on a lower-school history paper, an A-level paper, or as a PhD topic. The answers would (or should) differ according to the level. On mathematics examination papers, the question has to be tailored to the level in order to discriminate between the candidates: if it is too easy, nearly all candidates will score very high marks; if it is too hard, nearly all candidates will make little progress on any of the questions.

Setting STEP

STEP is produced under the auspices of the Cambridge Assessment examining board. The setting procedure starts 30 months before the date of the examination, when the three examiners (one for each paper, from schools or universities) are asked to produce a draft paper. The first drafts are then vetted by the STEP coordinator (me!), who tries to enforce uniformity of difficulty, checks suitability of material and style, and tries to reduce overlap between the papers. Examiners then produce a second draft, based on the coordinator's suggestions. The second drafts are agreed with the coordinator and then circulated to three moderators (normally school teachers), and to the other examiners, who produce written comments and discuss the drafts in a two-day meeting. The examiners then produce third drafts, taking into account the consensus at the meetings. These drafts are sent to a vetter, who works through the papers, pointing out mistakes and infelicities. The resulting draft is checked by a second vetter and finally by a team of students. At each stage, the drafts are produced camera-ready, using a special mathematical word-processing package called LaTeX (which is also used to typeset this book).[3]

STEP Questions

STEP questions do not fall into any one category. Typically, there will be a range of types on each of the papers. Here are some thoughts, in no particular order.

- My favourite sort of question is in two (or maybe more) parts: in the first part, candidates are asked to perform some unfamiliar task and are told how to do it (integration using a given substitution, or expressing a quartic as the algebraic sum of two squares, for example); for the later parts, candidates are expected to demonstrate that they have understood and learned from the first part by applying the method to a new and perhaps more complicated task.

- Another favourite of mine is the question which has different answers according to the value of a certain number (or *parameter*). A common example involves sketching a graph whose shape depends on whether a parameter is positive or negative. Ideally, the different values of the parameter are not given in the question, and candidate has to identify them for herself or himself.

---

[3] It is freely available, so you might like to try it out. You will probably find it good fun to use, but quite time-consuming and not really suitable for writing out the solutions of STEP problems.

- Another good type of question requires candidates to do some preliminary special-case work and then prove a general result.

- In another type, candidates have to show that they can understand and use new notation or a new theorem.

- Questions with several unrelated parts (for example, three integrals using different techniques) are generally avoided; but if they occur, there tends to be a 'sting in the tail' involving putting all the parts together in some way.

- Some questions do not rely on any part of the syllabus: instead they might require 'common sense', involving counting or seeing patterns, or they might involve some aspect of more elementary mathematics with an unusual slant. Such questions try to test capacity for clear and logical thought without using much mathematical knowledge (like the calendar question mentioned in the section on preparation below, or questions concerning islands populated by toads 'who always tell the truth' and frogs 'who always fib').

- Some questions are devised to check that you do not simply apply routine methods blindly. For example, a function might have a maximum value at the end of the interval upon which it is defined, even though its derivative might be non-zero there. Finding the maximum in such a case is not simply a question of routine differentiation.

- There are always questions specifically on integration or differentiation, and many others (including mechanics and probability) that use calculus as a means to an end.

- Graph-sketching is regarded by mathematicians as a fundamental skill and there are nearly always questions that require a sketch.

- Basic ideas from analysis, such as

$$0 \leqslant \mathrm{f}(x) \leqslant k \Rightarrow 0 \leqslant \int_a^b \mathrm{f}(x)\mathrm{d}x \leqslant (b-a)k$$

    or

$$\mathrm{f}'(x) > 0 \Rightarrow \mathrm{f}(b) > \mathrm{f}(a) \text{ for } b > a$$

    or the relationship between an integral and a sum often come up, though knowledge of such results is never assumed — candidates may be told that they can use the result without proof, or a sketch 'proof' may be asked for.

- As mentioned above, questions on statistical tests are rare, because questions that require real understanding (rather than 'cookbook' methods) tend to be too difficult. More often, the questions in the Probability and Statistics section are about probability.

- The mechanics questions normally require a firm understanding of the basic principles (when to apply conservation of momentum and energy, for example) and may well involve a differential equation. Projectile questions are often set, but are never routine.

Advice to candidates

First appearances

I am often asked whether STEP is 'difficult'. Of course, it depends on what is meant by 'difficult'; it is not difficult compared with the mathematics I do every day. But to be on the safe side, I always answer 'yes' before explaining further.

Your first impression on looking at a STEP paper is likely to be that it does indeed look very difficult. Don't be discouraged! Its difficult appearance is largely due to it being very different in style from what you are used to.

At the time of writing, a typical A-level examination lasts 90 minutes and contains 10 compulsory questions. That is 9 minutes per question. If you are considering studying mathematics at a top university, it is likely that you will manage to do them all and get them nearly all right in the time available. A STEP examination lasts 3 hours, and you are only supposed to do six questions in three hours. You are very likely to get a grade 1 if you manage four questions (not necessarily complete); that means that each question is designed to take 45 minutes. If you compare a 9 minute question with a 45 minute question, of course the 45 minute question looks very hard!

You may be put off by the number of subjects covered on the paper. You should not be. STEP is supposed to provide sufficient questions for all candidates, no matter which mathematics syllabus (at the appropriate level) they have covered. It would be a very exceptional candidate who had the knowledge required to do all the questions. And there is plenty of choice (6 questions out of 13).

Once you get used to the idea that STEP is very different from A-level, it becomes much less daunting.

Preparation

The best preparation for STEP (apart, of course, from working through this excellent book) is to work slowly through old papers.[4] Hints and answers are available for some years, but you should use these with discretion: doing a question with hints and answers in front of you is nothing like doing it yourself, and you may well miss the whole point of the question (which is to make you think about mathematics). In general, thinking about the problem is much more important than getting the answer.

Should you try to learn up areas of mathematics that are not in your syllabus in preparation for STEP? The important thing to know is that it is much better to be very good at your syllabus than to have have a sketchy knowledge of lots of additional topics: depth rather than breadth is what matters. It may conceivably be worth your while to round out your knowledge of a topic you have already studied to fit in with the STEP syllabus; it is probably not worth your while to learn a new topic for the purposes of the exam, though I can think of a couple of exceptions:

- Hyperbolic functions obey simple rules similar to trigonometric functions (in fact, they are the same functions in the complex plane). If you haven't come across them, you can easily master them in a short time and this will open the door to many questions that would otherwise have been inaccessible to you.

- de Moivre's theorem (relating to complex numbers) is also very straightforward, though questions requiring de Moivre are less common.

It is worth emphasising that there is no 'hidden agenda': a candidate who does two complete probability questions and two complete mechanics questions will obtain the same mark and grade as one who does four complete pure questions.

Just as the examiners have no hidden agenda concerning syllabus, so they have no hidden agenda concerning your method of answering the question. If you can get to the end of a question correctly you

---

[4] These and the other publications mentioned below are obtainable from the STEP web site www.stepmathematics.org.uk. You will also find other useful material on the NRICH web site; do google it now if you haven't already done so.

will get full marks whatever method you use.[5] Some years ago one of the questions asked candidates to find the day of the week of a given date (say, the 5th of June 1905). A candidate who simply counted backwards day by day from the date of the exam would have received full marks for that question (but would not have had time to do any other questions).

You may be worried that the examiners expect some mysterious thing called rigour. Do not worry: STEP is an exam for schools, not universities, and the examiners understand the difference. Nevertheless, it is extremely important that you present ideas clearly, and show working at all stages.

Presentation

You should set out your answer legibly and logically (don't scribble down the first thought that comes into your head) – this not only helps you to avoid silly mistakes but also signals to the examiner that you know what you are doing (which can be effective even if you haven't the foggiest idea what you are doing).

Examiners are not as concerned with neatness as you might fear. However if you receive complaints from your teachers that your answers are difficult to follow then you should listen.[6] Remember that more space usually means greater legibility. Try writing on alternate lines (this leaves a blank line for corrections).

Try to read your answers with a hostile eye. Have you made it clear when you have come to the end of a particular argument? Try underlining your conclusions. Have you explained what you are trying to do? For example, if a question asks 'Is A true?' try beginning your answer by writing 'A is true' — if you think that it is true — so that the examiner knows which way your argument leads. If you used some idea (for example, integration by substitution), did you tell the examiner that this is what you were doing?

What to do if you cannot get started on a problem

Try the following, in order.

- Reread the question to check that you understand what is wanted.

- Reread the question to look for clues – the way it is phrased, or the way a formula is written, or other relevant parts of the question. (You may think that the setters are trying to set difficult questions or to catch you out. Usually, nothing could be further from the truth: they are probably doing all in their power to make it easy for you by trying to tell you what to do).

- Try to work out exactly what it is that you don't understand.

- Simplify the notation – e.g. by writing out sums explicitly.

- Look at special cases (choose special values which simplify the problem) in order to try to understand why a result is true.

- Write down your thoughts – in particular, try to express the exact reason why you are stuck.

---

[5] Though you must obey instructions in the question: for example, if it says 'Hence prove ...', then you must use the previous result in your proof.

[6] Begin rant: I am very surprised at the scrappy and illegible work that I receive from a few of my students. It seems so disrespectful to expect me to spend ages trying to decipher their work when they could have spent a little more time making it presentable, for example by copying it out neatly or writing more slowly. Why is my time less important than theirs? End rant.

- Go on to another question and go back later.

- If you are preparing for the examination (but not in the actual examination!) take a short break.[7]

- Discuss it with a friend or teacher (again, better not do this in the actual examination) or consult the hints and answers, but make sure you still think it through yourself.

    BUT REMEMBER: following someone else's solution is not remotely the same thing as doing the problem yourself. Once you have seen someone else's solution to a problem, then you are deprived, for ever, of much of the benefit that could have come from working it out yourself.

    Even if, ultimately, you get stuck on a particular problem, you derive vastly more benefit from seeing a solution to something with which you have already struggled, than by simply following a solution to something to which you've given very little thought.

What if a problem isn't coming out?

If you have got started but the answer doesn't seem to be coming out, then try the following.

- Check your algebra. In particular, make sure that what you have written works in special cases. For example: if you have written the series for $\log(1+x)$ as
$$1 - x + \tfrac{1}{2}x^2 - \tfrac{1}{3}x^3 + \cdots$$
then a quick check will reveal that it doesn't work for $x = 0$; clearly, the 1 should not be there.[8]

    A note on the subject of algebra. In many of the problems in this book, the algebra is quite stiff: you have to go through many lines of calculation before you get to an expression recognisably close to your target. Really, the only way to manage this efficiently is to check each line carefully before going on the next line. Otherwise, you can waste hours.[9]

- Make sure that what you have written makes sense. For example, in a problem which is dimensionally consistent, you cannot add x (with dimension length, say) to x^2 or to $\exp x$ (which itself does not make sense — the argument of $\exp$ has to be dimensionless). Even if there are no dimensions in the problem, it is often possible to mentally assign dimensions and hence enable a quick check.

    Be wary of applying familiar processes to unfamiliar objects (very easy to do when you are feeling at sea): for example, it is all too easy, if you are not sure where your solution is going, to solve the vector equation $\mathbf{a}.\mathbf{x} = 1$ by dividing both sides by vector $\mathbf{a}$; a bad idea.

- Analyse exactly what you are being asked to do. Try to understand the hints, explicit and implicit. Remember to distinguish between terms such as explain/prove/define/etc. (There is essentially no difference between 'prove' and 'show': the former tends to be used in more formal situations, but if you are asked to 'show' something, a proper proof is required.)

- Remember that different parts of a question are often linked. There may be guidance in the notation and choice of names of variables in the question.

---

[7] J. E. Littlewood (1885–1977), distinguished Cambridge mathematician and author of the highly entertaining *Mathematicians Miscellany* (Cambridge University Press, 1986) used to work seven days a week until an experiment revealed that when he took Sundays off the good ideas had a way of coming on Mondays.

[8] Another check will reveal that for very small positive x, log is positive (since its argument is bigger than 1) whereas the series is negative, so there is clearly something else wrong.

[9] Which what I normally do; but I am still hoping to heed my own advice in the not too distant future.

- If you get irretrievably stuck in the exam, state in words what you are trying to do and move on (at A-level, you don't get credit for merely stating intentions, but STEP examiners are generally grateful for any sign of intelligent life).

What to do after completing a question

It is a natural instinct to consider that you have finished with a question once you have got to an answer. However this instinct should be resisted both from a general mathematical point of view and from the much narrower view of preparing for an examination. Instead, when you have completed a question you should stop for a few minutes and think about it. Here is a check list for you to run through.

- Look back over what you have done, checking that the arguments are correct and making sure that they work for any special cases you can think of. It is surprising how often a chain of completely spurious arguments and gross algebraic blunders leads to the given answer.

- Check that your answer is reasonable. For example if the answer is a probability p then you should check that $0 \leqslant p \leqslant 1$. If your answer depends on an integer n, does it behave as it should when $n \to \infty$? Is it dimensionally correct?

  If, in the exam, you find that your answer is not reasonable, but you don't have time to do anything about it, then write a brief phrase showing that you understand that your answer is unreasonable (e.g. 'This is wrong because mass must be positive').

- Check that you have used all the information given. In many ways the most artificial aspect of examination questions is that you are given exactly the amount of information required to answer the question. If your answer does not use everything you are given then either it is wrong or you are faced with a very unusual examination question.

- Check that you have understood the point of the question. It is, of course, the case that not all exam questions have a point, but many do. What idea did the examiners want you to have? Which techniques did they want you to demonstrate? Is the result of the question interesting in some way? Does it generalise? If you can see the point of the question would your working show the point to someone who did not know it in advance?

- Make sure that you are not unthinkingly applying mathematical tools which you do not fully understand.[10]

- It is good preparation for the examination to try to see how the problem fits into the wider context and see if there is a special point which it is intended to illustrate. You may need help with this.

- As preparation for the examination, make sure that you actually understand not only what you have done, but also why you have done it that way rather than some other way. This is particularly important if you have had to use a hint or solution.

---

[10] Mathematicians should feel as insulted as engineers by the following joke.

A mathematician, a physicist and an engineer enter a mathematics contest, the first task of which is to prove that all odd numbers are prime. The mathematician has an elegant argument: '1's a prime, 3's a prime, 5's a prime, 7's a prime. *Therefore*, by mathematical induction, all odd numbers are prime. It's the physicist's turn: '1's a prime, 3's a prime, 5's a prime, 7's a prime, 11's a prime, 13's a prime, so, to within experimental error, all odd numbers are prime.' The most straightforward proof is provided by the engineer: '1's a prime, 3's a prime, 5's a prime, 7's a prime, 9's a prime, 11's a prime ...'.

- In the examination, check that you have given the detail required. There often comes a point in a question where, if we could show that A implies B, then the result follows. If, after a lot of thought, you suddenly see that A does indeed imply B the natural thing to do is to write triumphantly 'But A implies B so the result follows'[11]. Unfortunately unscrupulous individuals (not you, of course) who have no idea why A should imply B (apart from the fact that it would complete the question) could, and do, write the exactly the same thing. Go back through the major points of the question making sure that you have not made any major unexplained leaps.

---

[11] There is an old anecdote about the distinguished Professor X. In the middle of a lecture she writes 'It is obvious that A', suddenly falls silent and after a few minutes she rubs it out and walks out of the room. The awed students hear her pacing up and down outside. Then after twenty minutes she returns and writes 'It is obvious that A' and continues the lecture.

Worked Problems

Worked problem 1 (✓)

> Let
> $$f(x) = ax - \frac{x^3}{1+x^2},$$
> where a is a constant.
> Show that, if $a \geqslant 9/8$, then
> $$f'(x) \geqslant 0$$
> for all x.

2000 Paper I

First thoughts

When I play tennis and I see a ball that I think I can hit, I rush up to it and smack it into the net. This is a tendency I try to overcome when I am doing mathematics. In this question, for example, even though I'm pretty sure that I can find $f'(x)$, I'm going to pause for a moment before I do so. I am going to use the pause to think about two things.

- I'm going to think about what to do when I have found $f'(x)$.
- I'm going to try to decide what the question is really about.

Of course, I may not be able to decide what to do with $f'(x)$ until I actually see what it looks like; and I may not be able to see what the question is really about until I have finished it; maybe not even then — some questions are not really about anything in particular. I am also going to use the pause to think a bit about the best way of performing the differentiation. Should I simplify first? Should I make some sort of substitution? Clearly, knowing how to tackle the rest of the question might guide me in deciding the best way to do the differentiation. Or it may turn out that the differentiation is fairly straightforward, so that it doesn't matter how I do it.

Two more points occur to me as I re-read the question.

- I notice that the inequalities are not strict (they are $\geqslant$ rather than $>$). Am I going to have to worry about the difference?

- I also notice that there is an 'If ... then', and I wonder if this is going to cause me trouble. I will need to be careful to get the implication the right way round. I mustn't try to prove that if $f'(x) \geqslant 0$ then $a \geqslant 9/8$.

  It will be interesting to see why the implication is only one way — why it is not an 'if and only if' question. It may be just 'if' because 'only if' isn't true; or it may be just 'if' because the examiners thought that the question was long enough without the 'only if'.

Don't turn over until you have spent a little time thinking along these lines.

Doing the question

Looking ahead, it is clear that the real hurdle is going to be showing that $f'(x) \geq 0$. How am I going to do that? Two ways suggest themselves. First, if $f'(x)$ turns out to be a quadratic function, or an obviously positive multiple of a quadratic function, I should be able to use some standard method: looking at the discriminant ('$b^2 - 4ac$'), etc; or, better, completing the square. But if $f'(x)$ is not of this form, I will have to think of something else: maybe I will have to sketch a graph. I'm hoping that I won't have to sketch a graph, because then I could be here all night.

I've just noticed that f is an odd function, i.e. $f(x) = -f(-x)$. (Did you notice that?) That is helpful, because it means that $f'(x)$ is an even function.[12] If it had been odd, there could have been a difficulty, because all odd functions (at least, those with no vertical asymptotes) cross the horizontal axis $y = 0$ at least once and cannot therefore be positive for all x.

Now, how should I do the differentiation? I could divide out the fraction, giving

$$\frac{x^3}{1+x^2} = x - \frac{x}{1+x^2}$$

and

$$f(x) = (a-1)x + \frac{x}{1+x^2} = bx + \frac{x}{1+x^2}, \qquad (\dagger)$$

where I have set $b = a - 1$ to save writing. (Is this right? I'll just check that it works when $x = 2$. Yes, it does: $f(2) = 2a - \frac{8}{5} = 2(a-1) + \frac{2}{5}$.) This might save a bit of writing, but I don't at the moment see it helping me towards a positive function. On balance, I think I'll stick with the original form.

Another thought: should I differentiate the fraction using the quotient formula or should I write it as $x^3(1+x^2)^{-1}$ and use the product rule? I doubt if there is much in it. I never normally use the quotient rule — it's just extra baggage to carry round. But on this occasion, since the final form I am looking for is a single fraction and the denominator using the quotient rule is a square and therefore non-negative, I will.

One more thought: since I am trying to obtain an inequality, I must be careful throughout not to cancel any quantity which might be negative; or at least if I do cancel a negative quantity, I must remember to reverse the inequality.

Here goes (at last):

$$f'(x) = a - \frac{3x^2(1+x^2) - 2x(x^3)}{(1+x^2)^2}$$
$$= \frac{a + (2a-3)x^2 + (a-1)x^4}{(1+x^2)^2}.$$

I had to do a bit of algebra to obtain the second equation.

This is working out as I had hoped: the denominator is certainly positive and the numerator is a quadratic function of x^2. I can finish this off most elegantly by completing the square. There are two ways of doing this. Just to be clear in my mind, I'm going to write the numerator as

$$A + Bx^2 + Cx^4,$$

where $A = a$, $B = 2a - 3$ and $C = a - 1$. Then I can complete the square in two ways:

$$A\bigl(1 + (B/2A)x^2)\bigr)^2 + (C - B^2/4A)x^4$$

---

[12] You can see this easily by sketching a 'typical' odd function, or by differentiating the Maclaurin expansion of an odd function.

or
$$C(x^2 + B/2C))^2 + (A - B^2/4C).\qquad(\ddagger)$$

It doesn't seem to matter which I use. The second expression ($\ddagger$) looks a bit simpler

If $a \geqslant \tfrac{9}{8}$, then certainly $C > 0$. That means that the first of the two terms of ($\ddagger$) is positive: $C(x^2 + B/2C)^2 > 0$.

The second term of ($\ddagger$) can be written in terms of a as follows:
$$\begin{aligned}A - B^2/4C &= a - \frac{(2a-3)^2}{4(a-1)} \\ &= \frac{4a(a-1) - (2a-3)^2}{4(a-1)} \\ &= \frac{8a-9}{4(a-1)}.\end{aligned}\qquad(*)$$

It is very pleasing to see the numerator $8a - 9$; it is exactly what I want, because it is non-negative if $a \geqslant \tfrac{9}{8}$. That is what I needed to complete the question.

Post-mortem

Now I can look back and analyse what I have done.

On the technical side, it seems I was right not to use the simplified form of $x^3/(1+x^2)$ given in ($\dagger$). This would have lead to the quadratic $(b+1) + (2b-1)x^2 + bx^4$, which isn't any easier to handle than the quadratic involving a and just gives an extra opportunity to make an algebraic error.

Although introducing new variables A, B and C seemed at first to complicate the problem, it was useful to have them when it came to completing the square, which would have been a bit of a mess had I worked directly from the quartic expression with a in it.

Actually, I see on re-reading my solution that I have made a bit of a meal out of the ending. I needn't have completed the square at all; I could used the inequality $a \geqslant 9/8$ immediately after finding $f'(x)$, since a appears in $f'(x)$ with a plus sign always:
$$f'(x) = \frac{a + (2a-3)x^2 + (a-1)x^4}{(1+x^2)^2} \geqslant \frac{\tfrac{9}{8} + (\tfrac{9}{4} - 3)x^2 + (\tfrac{9}{8} - 1)x^4}{(1+x^2)^2} = \frac{(x^2-3)^2}{8(1+x^2)^2} \geqslant 0$$

as required.

However, my unnecessarily elaborate proof, involving completing the square, makes the role of a a bit clearer than it is in the shorter alternative. In fact, I can see how to answer my question about 'if and only if', namely is it the case that $f'(x) \geqslant 0$ for all x if *and only if* $a \geqslant \tfrac{9}{8}$? The answer is Yes. Looking at ($\ddagger$), I see that if $f'(x) \geqslant 0$ for all x, then certainly $C \geqslant 0$, otherwise for large enough x the first (squared) term would be dominant and negative. But if $0 \leqslant C < \tfrac{1}{8}$, then $B < 0$ and there is a value of x^2 for which the squared term in ($\ddagger$) vanishes, leaving only the second term which is negative, as can be seen from ($*$). That means $f'(x) < 0$ for this value of x contradicting our assumption. Therefore, $f'(x) \geqslant 0$ for all x implies that $C \geqslant \tfrac{1}{8}$ and $a \geqslant \tfrac{9}{8}$.

I wonder why the examiner wanted me to investigate the sign of $f'(x)$. The obvious reason is to see what the graph looks like. We can now see what this question is about. It is clear that the examiners really wanted to set the question: 'Sketch the graphs of the function $ax - x^3/(1+x^2)$ in the different cases that arise according to the value of a' but it was thought too long or difficult. It is worth looking back over my working to see what can be said about the shape of the graph of $f(x)$ when $a < 9/8$.

(I leave that to you to think about!)

Worked problem 2 (✓)

> The n positive numbers $x_1, x_2, \ldots, x_n$, where $n \geqslant 3$, satisfy
>
> $$x_1 = 1 + \frac{1}{x_2}, \quad x_2 = 1 + \frac{1}{x_3}, \quad \ldots, \quad x_{n-1} = 1 + \frac{1}{x_n},$$
>
> and also
>
> $$x_n = 1 + \frac{1}{x_1}.$$
>
> Show that
>
> (i) $x_1, x_2, \ldots, x_n > 1$,
>
> (ii) $x_1 - x_2 = -\dfrac{x_2 - x_3}{x_2 x_3}$,
>
> (iii) $x_1 = x_2 = \cdots = x_n$.
>
> Hence find the value of x_1.

1999 Paper I

First thoughts

My first thought is that this question has an unknown number of variables: $x_1, \ldots, x_n$. That makes it seem rather complicated. I might, if necessary, try to understand the result by choosing an easy value for n (maybe $n = 3$). If I manage to prove some of the results in this special case, I will certainly go back to the general case: doing the special case might help me tackle the general case, but I don't expect to get many marks in an exam if I just prove the result in one special case.

Next, I see that the question has three sub-parts, then a final one. The final one begins 'Hence ...'. This means that I must use at least one of the previous parts in my working for the final part. It is not clear from the structure of the question whether the three sub-parts are independent; the proof of (ii) and (iii) may require the previous result(s), or it may not.

Actually, I think I can see how to do the very last part. If I assume that (iii) holds, so that $x_1 = x_2 = \cdots = x_n = x$ (say), then each of the equations given in the question is identical and each gives a simple equation for x.

It is surprising, isn't it, that the final result doesn't depend on the value of n? That makes the idea of choosing $n = 3$, just to see what is going on, quite attractive, but I'm not going to resort to that idea unless a get very stuck.

The notation in part (i) is a bit odd. I'm not sure that I have seen anything like it before.[13] But it can only mean that each of the variables $x_1, x_2, \ldots x_n$ is greater than 1.

In fact, now I think about it, I am puzzled about part (i). How can an *inequality* help to derive the *equalities* in the later parts? I can think of a couple of ways in which the result $x_i > 1$ could be used. One is that I may need to cancel, say, x_1 from both sides of an equation in which case I would need to know that $x_1 \neq 0$. But looking back at the question, I see I already know that $x_1 > 0$ (it is given right at the

---

[13] It is just the sort of thing that is used in university texts; but I'm not sure that it would be used in STEP papers nowadays.

start of the question), so this cannot be the right answer. Maybe I need to cancel some other factor, such as $(x_1 - 1)$. Another possibility is that I get two or more solutions by putting $x_1 = x_2 = \cdots = x_n = x$ and I need the one with $x > 1$. This may be the answer: looking back at the question again, I see that it asks for *the* value of x_1 — so I am looking for a single value. I'm still puzzled, but I will remember to keep a sharp look out for ways of using part (i).

One more thing strikes me about the question. The equations satisfied by the x_i are given on two lines ('and also'). This could be for typographic reasons (the equations would not all fit on one line) but more likely it is to make sure that I have noticed that the last equation is a bit different: all the other equations relate x_i to x_{i+1}, whereas the last equation relates x_n to x_1. It goes back to the beginning, completing the cycle. I'm pleased that I thought of this, because this circularity must be important.

Doing the question

I think I will do the very last part first, and see what happens.

Suppose, assuming the result of part (iii), that $x_1 = x_2 = \cdots = x_n = x$. Then substituting into any of the equations given in the question gives

$$x = 1 + \frac{1}{x}$$

i.e. $x^2 - x - 1 = 0$. Using the quadratic formula gives

$$x = \frac{1 \pm \sqrt{5}}{2},$$

which does indeed give two answers (despite the fact that the question asks for just one). However, I see that one is negative and can therefore be eliminated by the condition $x_i > 0$ which was given in the question (not, I note, the condition $x_1 > 1$ from part (i); I still have to find a use for this).

I needed only part (iii) to find x_1, so I expect that either I need both (i) and (ii) directly to prove (iii), or I need (i) to prove (ii), and (ii) to prove (iii).

Now that I have remembered that $x_i > 0$ for each i, I see that part (i) is obvious. Since $x_2 > 0$ then $1/x_2 > 0$ and the first equation given in the question, $x_1 = 1 + 1/x_2$, shows immediately that $x_1 > 1$ and the same applies to x_2, x_3, etc.

Now what about part (ii)? The given equation involves x_1, x_2 and x_3, so clearly I must use the first two equations given in the question:

$$x_1 = 1 + \frac{1}{x_2}, \quad x_2 = 1 + \frac{1}{x_3}.$$

Since I want $x_1 - x_2$, I will see what happens if I subtract the two equations:

$$x_1 - x_2 = (1 + \frac{1}{x_2}) - (1 + \frac{1}{x_3}) = \frac{1}{x_2} - \frac{1}{x_3}i = \frac{x_3 - x_2}{x_2 x_3} = -\frac{x_2 - x_3}{x_2 x_3}. \quad (*)$$

That seems to work!

One idea that I haven't used so far is what I earlier called the circularity of the equations: the way that x_n links back to x_1. I'll see what happens if I extend the above result. Since there is nothing special about x_1 and x_2, the same result must hold if I add 1 to each of the suffices:

$$x_2 - x_3 = \frac{x_3 - x_4}{x_3 x_4}.$$

I see that I can combine this with the previous result:

$$x_1 - x_2 = -\frac{x_2 - x_3}{x_2 x_3} = \frac{x_3 - x_4}{x_2 x_3^2 x_4}.$$

I now see where this is going. The above step can be repeated to give

$$x_1 - x_2 = \frac{x_3 - x_4}{x_2 x_3^2 x_4} = -\frac{x_4 - x_5}{x_2 x_3^2 x_4^2 x_5} = \cdots$$

and eventually I will get back to $x_1 - x_2$:

$$x_1 - x_2 = -\frac{x_4 - x_5}{x_2 x_3^2 x_4^2 x_5} = \cdots = \pm \frac{x_1 - x_2}{x_2 x_3^2 x_4^2 x_5^2 \ldots x_n^2 x_1^2 x_2}$$

i.e.

$$(x_1 - x_2)\left(1 \mp \frac{1}{x_1^2 x_2^2 x_3^2 x_4^2 x_5^2 \ldots x_n^2}\right) = 0 \,.$$

I have put in a $\pm$ because each step introduces a minus sign and I'm not sure yet whether the final sign should be $(-1)^n$ or $(-1)^{n-1}$. I can check this later (for example, by working out one simple case such as $n = 3$); but I may not need to.

I deduce from this last equation that either

$$x_1^2 x_2^2 x_3^2 x_4^2 x_5^2 \ldots x_n^2 = \pm 1$$

or

$$x_1 = x_2$$

(which is what I want). At last I see where to use part (i): I know that

$$x_1^2 x_2^2 x_3^2 x_4^2 x_5^2 \ldots x_n^2 \neq \pm 1$$

because $x_1 > 1$, $x_2 > 1$, etc. Thus the only possibility is $x_1 = x_2$. Since there was nothing special about x_1 and x_2, I deduce further that $x_2 = x_3$, and so on, as required.

Post-mortem

There were a number of useful points in this question.

1. The first point concerns using the information given in the question. The process of teasing information from what is given is fundamental to the whole of mathematics. It is very important to study what is given (especially seemingly unimportant conditions, such as $x_i > 0$) to see why they have been given. If you find you reach the end of a question without apparently using some given information, then you should look back over your work: it is very unlikely that a condition has been given that is not used in some way. It may not be a necessary condition — and we will see that the condition $x_i > 0$ is not, in a sense, necessary in this question — but it should be sufficient. The other piece of information in the question which you might easily have overlooked is the use of the singular rather than the plural in referring to the solution (' ... find *the value* of ...'), implying that there is just one value, despite the fact that the final equation is quadratic.

2. The second point concerns the structure of the question. Here, the position of the word 'Hence' suggested strongly that none of the separate parts were stand-alone results; each had to be used for a later proof. Understanding this point made the question much easier, because I was always on the look out for an opportunity to use the earlier parts. Of course, in some problems (without that 'hence') some parts may be stand-alone; though this is rare in STEP questions.

   You may think that this is like playing a game according to hidden (STEP) rules, but that is not the case. Precision writing and precision reading is vital in mathematics and in many professions (law, for example). Mathematicians have to be good at it, which is the reason why so many employers want to recruit people with mathematical training.

3. The third point was the rather inconclusive speculation about the way inequalities might help to derive an equality. It turned out that what was actually required was $x_1 x_2 \ldots x_n \ne \pm 1$. I was a bit puzzled by this possibility in my first thoughts, because it seemed that the result ought to hold under conditions different from those given; for example, $x_i < 0$ for all i (does this condition work??). Come to think of it, why are conditions given on all x_i when they are all related by the given equations? This makes me think that there ought to be a better way of proving the result which would reveal exactly the conditions under which it holds.

4. Then there was the idea (which I didn't actually use) that I might try to prove the result for, say, $n = 3$ to help me understand what was going on. This would not have counted as a proof of the result (or anything like it), but it might have given me ideas for tackling the question.

5. A key observation was that the equations given in the question are 'circular'. It was clear that the circularity was essential to the question and it turned out to be the key to the most difficult part. Having identified it early on, I was ready to use it when the opportunity arose.

6. Finally, I was pleased that I read the whole question carefully before plunging in. This allowed me to see that I could easily do the last part before the preceding parts, which I found very helpful in getting into the question.

Final thoughts

It occurs to me only now, after my post-mortem, that there is another way of obtaining the final result.

Suppose I start with the idea of circularity (as indeed I might have, had I not been otherwise directed by the question) and use the given equations to find x_1 in terms of first x_2, then x_3, then x_4 and eventually in terms of x_1 itself. That should give me an equation I can solve, and I should be able to find out what conditions are needed on the x_i. Try it. You may need to guess a formula for x_1 in terms of x_i from a few special cases, then prove it by induction.[14] You will find it useful to define a sequence of numbers F_i such that $F_0 = F_1 = 1$ and $F_{n+1} = F_n + F_{n-1}$. (These numbers are called *Fibonacci numbers*.[15]) You should find that if $x_n = x_1$ for any n (greater than 1), then $x_n = \frac{1 \pm \sqrt{5}}{2}$.

---

[14] I went as far as x_7 to be sure of the pattern: I found that $x_7 = \frac{8x_1 + 5}{5x_1 + 3}$.

[15] Fibonacci (short for *filius* Bonacci — son of Bonacci) was called the greatest European mathematician of the Middle Ages. He was born in Pisa (Italy) in about 1175 AD. He introduced the series of numbers named after him in his book of 1202 called *Liber Abbaci* (*Book of the Abacus*). It was the solution to the problem of the number of pairs of rabbits produced by an initial pair: *A pair of rabbits are put in a field and, if rabbits take a month to become mature and then produce a new pair every month after that, how many pairs will there be in twelve months time?*

Problems

Problem 1: An integer equation (✓)

(i) Find all sets of positive integers a, b and c that satisfy the equation

$$\frac{1}{a} + \frac{1}{b} + \frac{1}{c} = 1.$$

(ii) Determine the sets of positive integers a, b and c that satisfy the inequality

$$\frac{1}{a} + \frac{1}{b} + \frac{1}{c} \geq 1.$$

1993 Paper I

Comments

Age has not diminished the value of this old chestnut. It requires almost no mathematics in the sense of examination syllabuses, but instead tests a vital asset for a mathematician, namely the capacity for systematic thought. For this reason, it tends to crop up in mathematics contests, where competitors come from different backgrounds. I most recently saw it in the form 'Find all the positive integer solutions of $bc + ca + ab = abc$'. (You see the connection?)

You will want to make use of the symmetry between a, b and c: if, for example, $a = 2$, $b = 3$ and $c = 4$, then the same solution can be expressed in five other ways, such as $a = 3$, $b = 4$ and $c = 2$. You do not want to derive all these separately, so you need to order a, b and c in some way.

The reason the question uses the word 'determine', rather than 'find' for part (ii) is that in part (i) you can write down all the possibilities explicitly, whereas for part (ii) there are, in some cases, an infinite number of possibilities which obviously cannot be explicitly listed, though they can be described and hence determined.

Solution to problem 1

Let us take, without any loss of generality, $a \leqslant b \leqslant c$.

(i) How small can a be?

First set $a = 1$. This gives no solutions, because it leaves nothing for $1/b + 1/c$.

Next set $a = 2$ and try values of b (with $b \geqslant 2$, since we have assumed that $a \leqslant b \leqslant c$) in order:
  if $b = 2$, then $1/c = 0$, which is no good;
  if $b = 3$, then $c = 6$, which works;
  if $b = 4$, then $c = 4$, which works;
  if $b \geqslant 5$, then $c \leqslant b$, so we need not consider this.

Then set $a = 3$, and try values of b (with $b \geqslant 3$) in order:
  if $b = 3$, then $c = 3$, which works;
  if $b \geqslant 4$, then $c \leqslant b$, so we need not consider this.

Finally, if $a \geqslant 4$, then at least one of b and c must be $\leqslant a$, so we need look no further.

The only possibilities are therefore $(2, 3, 6)$, $(2, 4, 4)$ and $(3, 3, 3)$.

(ii) Clearly, we must include all the solutions found in part (i). We proceed systematically, as in part (i).

First set $a = 1$. This time, any values of b and c will do (though we decided to choose $b \leqslant c$).

Next set $a = 2$ and try values of b (with $b \leqslant 2$, since we have assumed that $a \leqslant b \leqslant c$) in order:
  if $b = 2$, then any value of c will do (with $b \leqslant c$).
  if $b = 3$, then 3, 4, 5 and 6 will do for c, but 7 is too big;
  if $b = 4$, then $c = 4$ will do, but 5 is too big for c;
  if $b \geqslant 5$, then $c \leqslant b$, so we need not consider this.

Then set $a = 3$ and try values of b (with $b \geqslant 3$) in order:
  if $b = 3$, then $c = 3$, which works, but $c = 4$ is too big;
  if $b \geqslant 4$, then $c \leqslant b$, so we need not consider this.

As before $a = 4$ and $b \geqslant 4$, $c \geqslant 4$ gives no possibilities.

Therefore the extra sets for part (ii) are of the form $(1, b, c)$, $(2, 2, c)$, $(2, 3, 3)$, $(2, 3, 4)$ and $(2, 3, 5)$.

Post-mortem

A slightly different approach for part (i), which would also generalise for part (ii), is to start with the case $a = b = c$, for which the only solution is $a = b = c = \frac{1}{3}$. If a, b and c are not all equal then one of them, which we may take to be a, must be greater than $\frac{1}{3}$, i.e. $\frac{1}{2}$. The two remaining possibilities follow easily.

As long as you are methodical, it doesn't matter how you approach the question.

Having done the question, you might well want to investigate whether it generalises: could we replace the 1 on the right hand side with 2? Could we have four reciprocals instead of three? After a few scribbles, I decided that these other cases are not very interesting; but you might find something I missed.

Problem 2: Partitions of 10 and 20

> (i) Show that you can make up 10 pence in eleven ways using 10p, 5p, 2p and 1p coins.
>
> (ii) In how many ways can you make up 20 pence using 20p, 10p, 5p, 2p and 1p coins?

1997 Paper I

Comments

I don't really approve of this sort of question, at least as far as STEP is concerned, but I thought I'd better include one in this collection. The one given above seems to me to be a particularly bad example, because there are a number of neat and elegant mathematical ways of approaching it, none of which turn out to be any use.

The quickest instrument is the bluntest: just write out all the possibilities. Two things are important. First you must be systematic or you will get hopelessly confused; second, you must lay out your solution, with careful explanations, in a way which allows other people (examiners, for example) to understand exactly what you are doing. You can probably lay it out using pen and paper much better than I have overleaf, using a wordprocessing package. If you get an answer and, looking back, find that your work lacks clarity, then do it again.

Since different parts of STEP questions are nearly always related, you might be led to believe that the result of the second part follows from the first: you divide the required twenty pence into two tens and then use the result of the first part to give the number of ways of making up each 10. This would give an answer of 66 (why?) plus one for a single 20p piece. This would be neat, but the true answer is less than 67 because some arrangements are counted twice by this method — and it is not easy to work out which ones.

Solution to problem 2

Probably the best approach is to start counting with the arrangements which use as many high denomination coins as possible, then work down.

(i) We can make up 10p as follows:

10 (one way using only one 10p coin);

5+5 (one way using two 5p coins and no 10p coins);

5+2+2+1, 5+2+1+1+1, 5+1+1+1+1+1 (three ways using one 5p coin and no 10p coins);

2+2+2+2+2, 2+2+2+2+1+1, etc (six ways using no 5p or 10p coins).

That makes 11 ways altogether.

(ii) We can make up 20p as follows:

20 (one way using only a 20p coin);

10 + any of the 11 arrangements in the first part of the question (11 ways using one or two 10p coins);

5+5+5+5 (one way using four 5p coins and no 10p or 20p coins);

5+5+5+2+2+1, etc (3 ways using three 5p coins and no 10p or 20p coins);

5+5+2+2+2+2+2, 5+5+2+2+2+2+1+1, etc (6 ways using two 5p coins and no 10p or 20p coins);

5+2+2+2+2+2+2+1, etc (8 ways using one 5p coin and no 10p or 20p coins);

2+2+2+2+2+2+2+2+2+2, etc (11 ways using no 5p, 10p or 20p coins).

That makes 41 ways altogether.

Post-mortem

As in the previous question, the most important lesson to be learnt here is the value of a systematic approach and clear explanations. You should not be happy just to obtain the answer: there is no virtue in that. You should only be satisfied if you displayed your working at least as systematically as I have, above.

On reflection, this question is not quite as bad as I thought. I did in fact use the first part to help with the second. And in the second part I certainly used the method of setting out the different ways systematically that I developed in the less complicated first part.

Here is an interesting thought. For the part (i), consider the expansion of

$$\frac{1}{(1-x^{10})(1-x^5)(1-x^2)(1-x)} \quad (*)$$

in powers of x. This we can obtain from the binomial expansion of each of the four terms in the denominator, giving $(1+x^{10}+x^{20}+\cdots)(1+x^5+x^{10}+\cdots)(1+x^2+x^4+\cdots)(1+x+x^2+\cdots)$. Now if we multiply out these brackets and find that the term in x^{10} (say) is the sum of terms of the form $x^{10a} \times x^{5b} \times x^{2c} \times x^d$, one from each bracket, where a, b, c and d are non-negative integers such that $10a + 5b + 2c + d = 10$. It is not hard to see that there is exactly one possibility for these integers for each of the arrangements of the coins in part (i), so the number of arrangements is exactly equal to the coefficient of x^{10} in $(*)$.

Although this is neat, it doesn't help, because there is no easy way of obtaining the required coefficient. Formally, though, we could obtain the coefficient using Taylor series, which involves differentiating $(*)$ 10 times and setting $x = 0$. This is interesting, because we have converted a problem in discrete mathematics, involving only integers, to a problem in calculus involving only smooth functions.

Problem 3: Mathematical deduction (✓)

(i) Write down the average of the integers $n_1, (n_1+1), \ldots, (n_2-1), n_2$. Show that

$$n_1 + (n_1+1) + \cdots + (n_2-1) + n_2 = \tfrac{1}{2}(n_2 - n_1 + 1)(n_1 + n_2).$$

(ii) Write down and prove a general law of which the following are special cases:

$$\begin{aligned} 1 &= 0 + 1 \\ 2 + 3 + 4 &= 1 + 8 \\ 5 + 6 + 7 + 8 + 9 &= 8 + 27 \\ 10 + 11 + \cdots + 16 &= 27 + 64. \end{aligned}$$

Hence prove that

$$1^3 + 2^3 + 3^3 + \cdots + n^3 = \tfrac{1}{4}n^2(n+1)^2.$$

Comments

You could use induction to prove your general law but it is not necessary. The obvious alternative involves use of part (i). You might think of induction for the second bit of part (ii), but the question specifically tells you to use the previous result.

There are various ways of summing kth powers of integers, besides the trick given here, which does not readily generalise to powers other than the third. A simple way is to assume that the result is a polynomial of degree $k + 1$. You can then find the coefficients by substituting in $k + 1$ values of n to obtain $k+1$ simultaneous equations for the coefficients of the polynomial. We can guess that the leading coefficient will in general be $(k + 1)^{-1}$, because the sum approximates the area under the graph $y = x^k$ from $x = 1$ to $x = n$, which is given by the integral of x^k.

Solution to problem 3

(i) The average is $\frac{1}{2}(n_1 + n_2)$. (This is the mid-point of a ruler with n_1 at one end and n_2 at the other.) The sum of any numbers is the average of the numbers times the number of numbers, as given.

(ii) A general rule is
$$\sum_{k=m^2+1}^{(m+1)^2} k = m^3 + (m+1)^3.$$

We can prove this result by applying the formula derived in the first part of the question:
$$\sum_{k=m^2+1}^{(m+1)^2} k = \tfrac{1}{2}\big[(m+1)^2 - m^2\big]\big[(m+1)^2 + (m^2+1)\big]$$
$$= \tfrac{1}{2}\big[2m+1\big]\big[2m^2 + 2m + 2\big]$$
$$= 2m^3 + 3m^2 + 3m + 1$$
$$= m^3 + (m+1)^3.$$

For the last part, we can obtain sums of the cubes by adding together the general law for consecutive values of m:
$$1 + (2+3+4) + (5+6+7+8+9) + \cdots + \big((N-1)^2 + 1\big) + \cdots + N^2\big)$$
$$= (0+1) + (1+8) + (8+27) + \cdots + \big((N-1)^3 + N^3\big).$$

The left hand side is just the sum of integers from 1 to N^3, so using the first part gives
$$\tfrac{1}{2}N^2(N^2+1) = 2(1^3 + 2^3 + \cdots + N^3) - N^3$$

i.e.
$$\tfrac{1}{2}N^2(N^2 + 2N + 1) = 2(1^3 + 2^3 + \cdots + N^3)$$

as required.

Post-mortem

To find the general rule requires an understanding of the algebra you have just done: why does it work in the way it does?

Of course, there are many other 'general rules' besides the one given in the solution. The situation is similar to the 'find the next number in the sequence' questions which come up in IQ tests. As no lesser figure than the philosopher Wittgenstein has pointed out, there is no correct answer. Given any finite sequence of numbers, a formula can always be found which will fit all the given numbers and which makes the next number (e.g.) 42.

Nevertheless, when the above question was set, almost everyone produced the same general result (or no result at all). Mathematicians seem to know what sort of thing is required. Electronic computers might not have the faintest idea what to do, though they would no doubt complete the algebra rapidly.

Problem 4: Divisibility (✓)

(i) How many integers greater than or equal to zero and less than 1000 are not divisible by 2 or 5? What is the average value of these integers?

(ii) How many integers greater than or equal to zero and less than 9261 are not divisible by 3 or 7? What is the average value of these integers?

1999 Paper I

Comments

There are a number of different ways of tackling this problem, but it should be clear that whatever way you choose for part (i) will also work for part (ii) (especially when you realise the significance of the number 9261). A key idea for part (i) is to think in terms of blocks of 10 numbers, realising that all blocks of 10 are the same for the purposes of the problem.

Solution to problem 4

(i) Only integers ending in 1, 3, 7, or 9 are not divisible by 2 or 5. This is 4/10 of the possible integers, so the total number of such integers is 4/10 of 1,000, i.e. 400.

The integers can be added in pairs:

$$\text{sum} = (1 + 999) + (3 + 997) + \cdots + (499 + 501).$$

There are 200 such pairs, so the sum is $1,000 \times 200$ and the average is 500.

A simpler argument to obtain the average would be to say that this is obvious by symmetry: there is nothing in the problem that favours an answer greater (or smaller) than 500.

Alternatively, we can find the number of integers divisible by both 2 and 5 by adding the number divisible by 2 (i.e. 500) to the number divisible by 5 (i.e. 200), and subtracting the number divisible by both 2 and 5 (i.e. 100) since these have been counted twice. To find the sum, we can sum those divisible by 2 (using the formula for a geometric progression), add the sum of those divisible by 5 and subtract the sum of those divisible by 10.

(ii) Either of the above methods will work. In the first method you consider integers in blocks of 21 (essentially arithmetic to base 21): there are 12 integers in each such block that are not divisible by 3 or 7 (namely 1, 2, 4, 5, 8, 10, 11, 13, 16, 17, 19, 20) so the total number is $9261 \times 12/21 = 5292$. The average is $9261/2$ as can be seen using the pairing argument $(1+9260)+(2+9259)+\cdots$ or the symmetry argument.

Post-mortem

I was in year 7 at school when I had my first encounter with lateral thinking in mathematics. The problem was to work out how many houses a postman delivers to in a street of houses numbered from 1 to 1000, given that he or she refuses to deliver to houses with the digit 9 in the number. It seemed impossible to do it systematically, until the idea of counting in base 9 occurred; and then it seemed brilliantly simple. I didn't know I was counting in base 9; in those days, school mathematics was very traditional and base 9 would have been thought very advanced. All that was required was the idea of working out how many numbers can be made from nine digits (i.e. 9^3) instead of from ten digits (i.e. 10^3); and of course it doesn't matter which nine.

Did you spot the significance of the number 9261? It is 21^3, i.e. the number in base 10 that is written as 1000 in base 21. Note how carefully the question is written and laid out to suggest a connection between the two paragraphs (and also to highlight the difference). For the examination, the number 1,000,000 was used in part (i) instead of $1,000$ in order to discourage candidates from spending the first hour of the examination writing down the numbers from 1 to 1000.

Having realised that part (i) involves $(2 \times 5)^3$ and part (ii) involves $(3 \times 7)^3$ you are probably now wondering what the general result is, i.e:

How many integers greater than or equal to zero and less than $(pq)^3$ are not divisible by p or q? What is the average value of these integers? Let's make p and q co-prime, so that they have no common factors.

I leave this to you, with the hint that you need to think about only pq, rather than $(pq)^3$, to start with.

Problem 5: The modulus function (✓)

Find all the solutions of the equation

$$|x+1| - |x| + 3|x-1| - 2|x-2| = x+2 \,.$$

1997 Paper I

Comments

This looks more difficult than it is.

There is no clever way to deal with the modulus function: you have to look at the different cases individually. For example, for $|x|$ you have to look at $x \leqslant 0$ and $x \geqslant 0$. The most straightforward approach would be to solve the equation in each of the different regions determined by the modulus signs: $x \leqslant -1$, $-1 \leqslant x \leqslant 0$, $0 \leqslant x \leqslant 1$, etc. You might find a graphical approach helps you to picture what is going on (I didn't).

Solution to problem 5

Let
$$f(x) = |x+1| - |x| + 3|x-1| - 2|x-2| - (x+2).$$

We have to solve $f(x) = 0$ in the five regions of the x-axis determined by the modulus functions, namely

$$-\infty < x \leqslant -1; \quad -1 \leqslant x \leqslant 0; \quad 0 \leqslant x \leqslant 1; \quad 1 \leqslant x \leqslant 2; \quad 2 \leqslant x < \infty.$$

In the separate regions, we have

$$f(x) = \begin{cases} -(x+1) + x - 3(x-1) + 2(x-2) - (x+2) & = & -2x - 4 & \text{for} & -\infty < x \leqslant -1 \\ (x+1) + x - 3(x-1) + 2(x-2) - (x+2) & = & -2 & \text{for} & -1 \leqslant x \leqslant 0 \\ (x+1) - x - 3(x-1) + 2(x-2) - (x+2) & = & -2x - 2 & \text{for} & 0 \leqslant x \leqslant 1 \\ (x+1) - x + 3(x-1) + 2(x-2) - (x+2) & = & 4x - 8 & \text{for} & 1 \leqslant x \leqslant 2 \\ (x+1) - x + 3(x-1) - 2(x-2) - (x+2) & = & 0 & \text{for} & 2 \leqslant x < \infty \end{cases}$$

Solving in each region gives:

1. $-\infty < x \leqslant -1$: here, $f(x) = 0$ only if $x = -2$. This is a solution since the point $x = -2$ lies in the region $x \leqslant -1$.

2. $-1 \leqslant x \leqslant 0$: here, $f(x) = -2 \ (\neq 0)$ so there is no solution.

3. $0 \leqslant x \leqslant 1$: here, $f(x) = 0$ only if $x = -1$. This is not a solution since the point $x = -1$ does not lie in the region $0 \leqslant x \leqslant 1$.

4. $1 \leqslant x \leqslant 2$: here, $f(x) = 0$ only if $x = 2$, which is a solution since $x = 2$ lies in the range $1 \leqslant x \leqslant 2$.

5. $2 \leqslant x < \infty$: here, $f(x) = 0$ identically (i.e. for all values of x), so the equation is satisfied for all x in the region.

Collecting these results together shows that the equation $f(x) = 0$ is satisfied only by $x = -2$ and by any x in the region $x \geqslant 2$.

Post-mortem

As mentioned before, this should not be found too difficult once you identify the different regions and consider each on its own. Some care is needed, though, and it would be sensible to go back and check that the solutions do indeed satisfy the original equation.

Two small points of technique:

- Since it was necessary to refer to the original equation quite a few times, I found it useful to define a function f so that the equation became $f(x) = 0$.

    Alternatively, you could label the equation with a (∗), say, when you first write it down and then later say 'Equation (∗) becomes ...'.

- It helped me to set out the different cases very clearly. I numbered them later but it might well have saved writing to number them at the beginning.

Problem 6: The regular Reuleaux heptagon (✓)

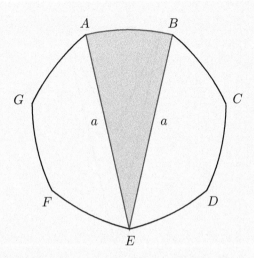

The diagram shows a British 50 pence coin. The seven arcs AB, BC, ..., FG, GA are of equal length and each arc is formed from the circle of radius a having its centre at the vertex diametrically opposite the mid-point of the arc. Show that the area of the face of the coin is

$$\frac{a^2}{2}\left(\pi - 7\tan\frac{\pi}{14}\right).$$

1987 Specimen Paper I

Comments

The first difficulty with this elegant problem is drawing the diagram. However, you can simplify both the drawing and the solution by restricting your attention to just one sector of a circle of radius a.

The figure sketched above has *constant diameter*; it can roll between two parallel lines without losing contact with either. (This looks plausible and you can verify it by sellotaping some 50 pence coins together, but a solid proof is not very easy.) The distance between these lines is the diameter of the figure. Like a circle, the 50 pence piece has circumference equal to π times the diameter, which is in fact always true for a figure with constant diameter.

Reuleaux polygons are general polygons of constant diameter, and a heptagon has seven sides like, for example, British 50 and 20 pence pieces, Botswanan 50 thebe coins and Jordanian half dinars. The shield on Lancia cars is a Reuleaux triangle.

Solution to problem 6

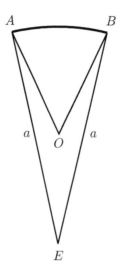

In the figure, the point O is equidistant from each of three vertices A, B and E. The plan is to find the area of the sector AOB by calculating the area of AEB and subtracting the areas of the two congruent isosceles triangles OBE and OAE. The required area is 7 times this.

First we need angle $\angle AEB$. We know that $\angle AOB = \tfrac{2}{7}\pi$ and hence $\angle BOE = \tfrac{1}{2}(2\pi - \tfrac{2}{7}\pi) = \tfrac{6}{7}\pi$ (using the sum of angles round the point O). Finally,

$$\angle AEB = 2\angle OEB = 2 \times \tfrac{1}{2}(\pi - \tfrac{6}{7}\pi) = \tfrac{1}{7}\pi,$$

using the sum of angles of an isosceles triangle.[16]

Now ABE is a sector of a circle of radius a, so its area is

$$\pi a^2 \times \frac{\tfrac{1}{7}\pi}{2\pi} = \frac{\pi a^2}{14}.$$

The area of triangle OBE is $\tfrac{1}{2}BE \times$ height, i.e.

$$\tfrac{1}{2}a \times \frac{a}{2}\tan\angle OBE = \frac{a^2}{4}\tan\frac{\pi}{14}.$$

The area of the coin is therefore

$$7 \times \left(\frac{\pi a^2}{14} - 2 \times \frac{a^2}{4}\tan\frac{\pi}{14}\right),$$

which reduces to the given answer.

Post-mortem

It is simple now to calculate the area of a regular n-sided Reuleaux polygon. You should of course find that the area tends to that of a circle of the same diameter as $n \to \infty$.

---

[16] After the first edition, a correspondent pointed out that the following argument gives angle EAB more quickly: clearly A, B and E all lie on a circle with centre O; the angle at O subtended by the chord AB is $\tfrac{2}{7}\pi$, so the angle at the circumference is $\tfrac{1}{7}\pi$. Obvious, really — can't think why I didn't see it.

Problem 7: Chain of equations (✓)

Suppose that
$$3 = \frac{2}{x_1} = x_1 + \frac{2}{x_2} = x_2 + \frac{2}{x_3} = x_3 + \frac{2}{x_4} = \cdots.$$

Guess an expression, in terms of n, for x_n. Then, by induction or otherwise, prove the correctness of your guess.

1997 Paper II

Comments

Wording this sort of question is a real headache for the examiners. Suppose you guess wrong; how can you then prove your guess by induction (unless you get that wrong too)? How else can the question be phrased? In the end, we decided to assume that you are all so clever that your guesses will all be correct.

To guess the formula, you need to work out x_1, x_2, x_3, etc and look for a pattern. You should not need to go beyond x_4.

Proof by induction is not in the core A-level syllabus. We decided to include it in the syllabus for STEP I and II because the idea behind it is not difficult and it is very important both as a method of proof and also as an introduction to more sophisticated mathematical thought.

Solution to problem 7

First let's put the equations into a more manageable form. Each equality can be written in the form

$$3 = x_n + \frac{2}{x_{n+1}}, \quad \text{i.e.} \quad x_{n+1} = \frac{2}{3 - x_n}.$$

We find $x_1 = \frac{2}{3}$, $x_2 = \frac{6}{7}$, $x_3 = \frac{14}{15}$ and $x_4 = \frac{30}{31}$. The denominators give the game away. We guess

$$x_n = \frac{2^{n+1} - 2}{2^{n+1} - 1}.$$

For the induction, we need a starting point: our guess certainly holds for $n = 1$ (and 2, 3, and 4!).

For the inductive step, we suppose our guess also holds for $n = k$, where k is any integer, so that

$$x_k = \frac{2^{k+1} - 2}{2^{k+1} - 1}.$$

If we can show that it then also holds for $n = k + 1$, we are done.

We have, from the equation given in the question,

$$x_{k+1} = \frac{2}{3 - x_k} = \frac{2}{3 - \frac{2^{k+1}-2}{2^{k+1}-1}} = \frac{2(2^{k+1} - 1)}{3(2^{k+1} - 1) - (2^{k+1} - 2)} = \frac{2^{k+2} - 2}{2^{k+2} - 1},$$

as required.

Post-mortem

There's not much to say about this. By STEP standards, it is fairly easy and short. Nevertheless, you are left to your own devices from the beginning, so you should be pleased if you got it out.

Perhaps the key step came in the very first line of the solution, when we had to decide how to separate out the equations. We could have tried instead

$$x_2 + \frac{2}{x_3} = x_3 + \frac{2}{x_4}$$

leading to

$$(x_2 - x_3) = \frac{2(x_3 - x_4)}{x_3 x_4},$$

and so on, but it would not be at all clear what to do next.

Problem 8: Trig. equations (✓✓)

(i) Show that, if $\tan^2 \theta = 2\tan\theta + 1$, then $\tan 2\theta = -1$.

(ii) Find all solutions of the equation
$$\tan\theta = 2 + \tan 3\theta$$
which satisfy $0 < \theta < 2\pi$, expressing your answers as rational multiples of π.

(iii) Find all solutions of the equation
$$\cot\phi = 2 + \cot 3\phi$$
which satisfy $-\frac{3\pi}{2} < \phi < \frac{\pi}{2}$, expressing your answers as rational multiples of π.

1997 Paper II

Comments

There are three distinct parts. It is pretty certain that they are related, but it is not obvious what the relationship is. Part (i) must surely help with part (ii) in some way that will only become apparent once part (ii) is under way.

In the absence of any other good ideas, it looks right to start part (ii) by expressing the double and triple angle tans in terms of single angle tans. You should remember the formula
$$\tan(A+B) = \frac{\tan A + \tan B}{1 - \tan A \tan B}.$$

You can use this for $\tan 3\theta$ and hence (later on) for $\cot 3\theta$. If you ever forget the $\tan(A+B)$ formula, you can quickly work it out from the corresponding sin and cos formulae:
$$\tan(A+B) = \frac{\sin(A+B)}{\cos(A+B)} = \frac{\sin A \cos B + \cos A \sin B}{\cos A \cos B - \sin A \sin B}.$$

You are are not expected to remember the more complicated triple angle formulae (I certainly don't).

You may well find yourself trying to solve cubic equations at some stage in this question; no need to panic — there is sure to be one easily spottable root, in which case you can reduce the cubic to a quadratic.

Interesting, isn't it, that the range of ϕ for part (iii) is not the obvious $0 < \theta < 2\pi$? Maybe that is significant.

Solution to problem 8

We will write t for $\tan\theta$ (or $\tan\phi$) throughout.

(i) $\tan 2\theta = \dfrac{2t}{1-t^2} = -1$ (since $2t = t^2 - 1$ by minor rearrangement of the given equation).

(ii) We first work out $\tan 3\theta$. We have

$$\tan 3\theta = \tan(\theta + 2\theta) = \frac{\tan\theta + \tan 2\theta}{1 - \tan\theta \tan 2\theta} = \frac{t + \frac{2t}{1-t^2}}{1 - t\frac{2t}{1-t^2}} = \frac{3t - t^3}{1 - 3t^2},$$

so the equation becomes

$$\frac{3t - t^3}{1 - 3t^2} = t - 2, \quad \text{i.e.} \quad t^3 - 3t^2 + t + 1 = 0.$$

One solution (by inspection) is $t = 1$. Thus one set of roots is given by $\theta = n\pi + \tfrac{1}{4}\pi$.

There are no other obvious integer roots, but we can reduce the cubic equation to a quadratic equation by dividing out the known factor $(t-1)$. I would start by writing $t^3 - 3t^2 + t + 1 \equiv (t-1)(t^2 + at - 1)$ since the coefficients of t^2 and of t^0 in the quadratic bracket are obvious. Then I would multiply out the brackets to find that $a = -2$. Now we see the connection with part (i): $t^2 + at - 1 = 0 \Rightarrow \tan 2\theta = -1$, and hence $2\theta = n\pi - \tfrac{1}{4}\pi$.

The roots are therefore $\theta = n\pi + \tfrac{1}{4}\pi$ and $\theta = \tfrac{1}{2}n\pi - \tfrac{1}{8}\pi$. The multiples of π in the given range are $\{\tfrac{1}{4}, \tfrac{3}{8}, \tfrac{7}{8}, \tfrac{5}{4}, \tfrac{11}{8}, \tfrac{15}{8}\}$.

(iii) For the last part, we could set $\cot\phi = \dfrac{1}{\tan\phi}$ and $\cot 3\phi = \dfrac{1}{\tan 3\phi}$ in the given equation, thereby obtaining

$$\frac{1}{t} = 2 + \frac{1 - 3t^2}{3t - t^3}.$$

This simplifies to the cubic equation $t^3 + t^2 - 3t + 1 = 0$. There is an integer root $t = 1$, and the remaining quadratic is $t^2 + 2t - 1 = 0$. Learning from the first part, we write this as $\dfrac{2t}{1-t^2} = 1$, which means that $\tan 2\phi = n\pi + \tfrac{1}{4}\pi$. Proceeding as before gives (noting the different range) the following multiples of π: $\{\tfrac{1}{4}, \tfrac{1}{8}, -\tfrac{3}{8}, -\tfrac{3}{4}, -\tfrac{7}{8}, -\tfrac{11}{8}\}$.

Post-mortem

There was a small but worthwhile notational point in this question: it is often possible to use the abbreviation t for tan (or s for sin, etc), which can save a great deal of writing.

There are two other points worth recalling. First is the way that part (i) fed into part (ii), but had to be mildly adapted for part (iii). This is a typical device used in STEP questions aimed to see how well you learn new ideas. Second is what to do when faced with a cubic equation. There is a formula for the roots of a cubic, but no one knows it nowadays. Instead, you have to find at least one root by inspection. Having found one root, you have a quick look to see if there are any other obvious roots and, if not, then divide out the known factor to obtain a quadratic equation.

The detectives amongst you will have worked out the reason for the peculiar choice $-\tfrac{3}{2}\pi < \phi < \tfrac{1}{2}\pi$ for part (iii). The reciprocal relation between tan and cot we used at the start of part (iii) is not the only way to relate these two trigonometric function. We could have instead used $\cot A = \tan(\tfrac{1}{2}\pi - A)$.

The equation of part (ii) transforms exactly into the equation of part (iii) if we set $\phi = \tfrac{1}{2}\pi - \theta$. Furthermore, the given range of ϕ corresponds exactly to the range of θ given in part (ii). We can therefore write down the solutions for part (iii) directly from the solutions for part (ii).

Problem 9: Integration by substitution (✓)

Show, by means of a change of variable or otherwise, that

$$\int_0^\infty \mathrm{f}((x^2+1)^{\frac{1}{2}} + x)\,\mathrm{d}x = \frac{1}{2}\int_1^\infty (1+t^{-2})\mathrm{f}(t)\,\mathrm{d}t\,,$$

for any given function f.
Hence, or otherwise, show that

$$\int_0^\infty \left((x^2+1)^{\frac{1}{2}} + x\right)^{-3}\,\mathrm{d}x = \frac{3}{8}\,.$$

1998 Paper I

Comments

Note that 'by change of variable' means the same as 'by substitution'.

There are two things to worry about when you are trying to find a change of variable to convert one integral to another: you need to make the integrands match up and you need to make the limits match up. Sometimes, the limits give the clue to the change of variable. (For example, if the limits on the original integral were 0 and 1 and the limits on the transformed integral were 0 and $\frac{1}{4}\pi$, then an obvious possibility would be to make the substitution $t = \tan x$). Here, the change of variable is determined by the integrand, since it must work for all choices of f.

Perhaps you are worried about the infinite upper limit of the integrals. If you are trying to prove some rigorous result about infinite integrals, you might use the definition

$$\int_0^\infty \mathrm{f}(x)\,\mathrm{d}x = \lim_{a\to\infty} \int_0^a \mathrm{f}(x)\,\mathrm{d}x\,,$$

but for present purposes you just do the integral and put in the limits. The infinite limit will not normally present problems. For example,

$$\int_1^\infty (x^{-2} + \mathrm{e}^{-x})\mathrm{d}x = \left(-x^{-1} - \mathrm{e}^{-x}\right)\Big|_1^\infty = -\frac{1}{\infty} - \mathrm{e}^{-\infty} + \frac{1}{1} + \mathrm{e}^{-1} = 1 + \mathrm{e}^{-1}\,.$$

Don't be afraid of writing $1/\infty = 0$. It is perfectly OK to use this as shorthand for $\lim\limits_{x\to\infty} 1/x = 0$; but ∞/∞, $\infty - \infty$ and $0/0$ are definitely not OK, because of their ambiguity.

Solution to problem 9

Clearly, to get the argument of f right, we must set

$$t = (x^2 + 1)^{\frac{1}{2}} + x \,.$$

We must check that the new limits are correct (if not, we are completely stuck). When $x = 0$, $t = 1$ as required. Also, $(x^2 + 1)^{\frac{1}{2}} \to \infty$ as $x \to \infty$, so $t \to \infty$ as $x \to \infty$. Thus the upper limit is still ∞, again as required.

Using the standard method of changing variable in an integral (basically the chain rule), the integral becomes

$$\int_1^\infty \mathrm{f}(t) \frac{\mathrm{d}x}{\mathrm{d}t}\, \mathrm{d}t$$

so the next task is to find $\dfrac{\mathrm{d}x}{\mathrm{d}t}$. This we can do in two ways: we find x in terms of t and differentiate it; or we could find $\dfrac{\mathrm{d}t}{\mathrm{d}x}$ and turn it upside down. The snag with the second method is that the answer will be in terms of x, so we will have to express x in terms of t anyway — in which case, we may as well use the first method.

We start by finding x in terms of t:

$$t = (x^2 + 1)^{\frac{1}{2}} + x \Rightarrow (t - x)^2 = (x^2 + 1) \Rightarrow x = \frac{t^2 - 1}{2t} = \frac{t}{2} - \frac{1}{2t} \,.$$

Then we differentiate:

$$\frac{\mathrm{d}x}{\mathrm{d}t} = \frac{1}{2} + \frac{1}{2t^2} = \frac{1}{2}(1 + t^{-2})$$

which is exactly the factor that we require for the transformed integrand given in the question.

For the last part, we take $\mathrm{f}(t) = t^{-3}$. Thus

$$\begin{aligned}
\int_0^\infty \left((x^2 + 1)^{\frac{1}{2}} + x\right)^{-3} \mathrm{d}x &= \frac{1}{2} \int_1^\infty (t^{-3} + t^{-5})\, \mathrm{d}t \\
&= \frac{1}{2} \left(\frac{t^{-2}}{-2} + \frac{t^{-4}}{-4} \right) \Big|_1^\infty \\
&= \frac{1}{2} \left(\frac{1}{2} + \frac{1}{4} \right) = \frac{3}{8} \,.
\end{aligned}$$

as required.

Post-mortem

This is not a difficult question conceptually once you realise the significance of the fact that the change of variable must work whatever function f is in the integrand.

There was a useful point connected with calculating $\dfrac{\mathrm{d}x}{\mathrm{d}t}$. It was a good idea not to plunge into the algebra without first thinking about alternative methods; in particular, we might have used (but didn't)

$$\frac{\mathrm{d}x}{\mathrm{d}t} = 1 \Big/ \frac{\mathrm{d}t}{\mathrm{d}x} \,.$$

Finally, there was the infinite limit of the integrand, which I hope you saw was not something to worry about (even though infinite limits are excluded from the A-level core). If you were setting up a formal definition of what an integral is, you would have to use finite limits, but if you are merely calculating the value of an integral, you just go ahead and do it, with whatever limits you are given.

Problem 10: True or false (✓✓)

Which of the following statements are true and which are false? Justify your answers.

(i) $a^{\ln b} = b^{\ln a}$ for all positive numbers a and b.

(ii) $\cos(\sin \theta) = \sin(\cos \theta)$ for all real θ.

(iii) There exists a polynomial P such that $|P(x) - \cos x| \leqslant 10^{-6}$ for all (real) x.

(iv) $x^4 + 3 + x^{-4} \geqslant 5$ for all $x > 0$.

1998 Paper I

Comments

The four parts are (somewhat annoyingly) related only by the fact that you have to decide whether each statement is true of false.

If true, the justification has to be a proof. If false, you could prove that it is false in some general way, but it is nearly always better to find a simple counterexample — as simple as possible.

Part (iii) might look a bit odd. I suppose it relates to the standard approximation $\cos x \approx 1 - \frac{1}{2}x^2$ which holds when x is small. This result can be improved by using a polynomial of higher degree: the next term is $+\frac{1}{4!}x^4$. It can be proved that $\cos x$ can be approximated as accurately as you like for small x by a polynomial of the form

$$\sum_{n=0}^{N}(-1)^n \frac{x^{2n}}{(2n)!}$$

(the truncated Maclaurin expansion). You have to use more terms of the approximation (i.e. a larger value of N) if you want either greater accuracy or larger x. In part (iii) of this question, you are being asked if there is a polynomial such that the approximation is good for all values of x.

I've awarded the question ✓✓ for difficulty, just because quite a few good ideas are required.

Solution to problem 10

(i) True. The easiest way to see this is to log both sides. For the left hand side, we have
$$\ln(a^{\ln b}) = (\ln b)(\ln a)$$
and for the right hand side we have
$$\ln(a^{\ln b}) = (\ln a)(\ln b),$$
which agree.

Note that we have to be a bit careful with this sort of argument. The argument used is that $A = B$ because $\ln A = \ln B$. This requires the property of the ln function that $\ln A = \ln B \Rightarrow A = B$. You can easily see that this property holds because ln is a strictly increasing function; if $A > B$, then $\ln A > \ln B$. The same would not hold for (say) sin (i.e. $\sin A = \sin B \not\Rightarrow A = B$).

(ii) False. $\theta = \frac{1}{2}\pi$ is an easy counterexample. Even though it is 'obvious' we still need to show that $\cos 1 \neq 0$, which we could do by noting that $0 < 1 < \frac{1}{3}\pi$ and sketching the graph of $\cos x$.

(iii) False. Roughly speaking, any polynomial can be made as large as you like by taking x to be very large (provided it is of degree greater than zero), whereas $|\cos x| \leq 1$; and there is obviously no polynomial of degree zero (i.e. no constant number) for which the statement holds.

But how can we write this out formally? First let us knock off the case when the polynomial is of degree zero, i.e. a constant, call it P. Then either $P \geq \frac{1}{2}$ or $P < \frac{1}{2}$. In either case, P cannot be close to both $\cos 0$ and $\cos \frac{1}{2}\pi$.

Now suppose
$$P(x) = a_N x^N + \sum_{n=0}^{N-1} a_n x^n \qquad (*)$$
where $N \geq 1$ and assume $a_N > 0$. It is enough to show that $P(x) > 2$ for some value of x. We can find a number x so large that $a_N x^N > N|a_n|x^n + 2$ for each integer n with $0 \leq n \leq N - 1$. The smallest possible value of $P(x)$ for any given positive x would be achieved if all the coefficients in the sum were negative. Thus
$$P(x) \geq a_N x^N - \sum_{n=0}^{N-1} |a_n| x^n > a_N x^N + (2 - a_N x^N) = 2.$$
and we are done.

(iv) True: $x^4 + 3 + x^{-4} = (x^2 - x^{-2})^2 + 5 \geq 5$.

Post-mortem

The important point here is that if you want to show a statement is true, you have to give a formal proof, whereas if you want to show that it is false, you only need give one counterexample. It does not have to be an elaborate counterexample — in fact, the simpler the better.

My proof for part (iii) is more elaborate than could have been expected of candidates in the examination. A sketch of a polynomial of degree N would have been enough, provided the special case of $N = 0$ was dealt with separately.

Problem 11: Egyptian fractions (✓)

A number of the form $\frac{1}{N}$, where N is an integer greater than 1, is called a *unit fraction*. Noting that
$$\frac{1}{2} = \frac{1}{3} + \frac{1}{6} \quad \text{and} \quad \frac{1}{3} = \frac{1}{4} + \frac{1}{12},$$
guess a general result of the form
$$\frac{1}{N} = \frac{1}{a} + \frac{1}{b} \qquad (*)$$
and hence prove that any unit fraction can be expressed as the sum of two distinct unit fractions.
By writing (*) in the form
$$(a - N)(b - N) = N^2$$
and by considering the factors of N^2, show that if N is prime, then there is only one way of expressing $\frac{1}{N}$ as the sum of two distinct unit fractions.

Prove similarly that any fraction of the form $\frac{2}{N}$, where N is prime number greater than 2, can be expressed uniquely as the sum of two distinct unit fractions.

2000 Paper II

Comments

Fractions written as the sum of unit fractions are called *Egyptian fractions*: they were used by Egyptians. The earliest record of such use is 1900BC. The Rhind papyrus in the British Museum gives a table of representations of fractions of the form $\frac{2}{n}$ as sums of unit fractions for all odd integers n between 5 and 101 — a remarkable achievement when you consider that algebra was 3,500 years in the future.

It is not clear why Egyptians represented fractions this way; maybe it just seemed a good idea at the time. Certainly the notation they used, in which for example $\frac{1}{n}$ was denoted by n with an oval on top, does not lend itself to generalisation to fractions that are not unit. One of the rules for expressing non-unit fractions in terms of unit fractions was that all the unit fractions in should be distinct, so $\frac{2}{7}$ had to be expressed as $\frac{1}{4} + \frac{1}{28}$ instead of as $\frac{1}{7} + \frac{1}{7}$, which seems pretty daft.

It is not obvious that every fraction can be express in Egyptian form; this was proved by Fibonacci in 1202. There are however still many unsolved problems relating to Egyptian fractions.

Egyptian fractions have been called a wrong turn in the history of mathematics; if so, it was a wrong turn that favoured style over utility; no bad thing, in my opinion.

Solution to problem 11

A good guess would be that the first term of the decomposition of $\dfrac{1}{N}$ is $\dfrac{1}{N+1}$, i.e. $a = N+1$. In that case, the other term is $\dfrac{1}{N} - \dfrac{1}{N+1}$ i.e. $b = N(N+1)$. That proves the result that every unit fraction can be expressed as the sum of two unit fractions.

The only factors of N^2 (since N is prime) are 1, N and N^2. The possible factorisations of N^2 are therefore $N^2 = 1 \times N^2$ or $N^2 = N \times N$. We discount $N \times N$, since this would lead to $a = b$, and this is ruled out in the question. That leaves only $a - N = 1$ and $b - N = N^2$ (or the other way round). Thus $N+1$ and $N^2 + N$ are the only possible values for a and b and the decomposition is unique.

For the second half, set
$$\frac{2}{N} = \frac{1}{a} + \frac{1}{b}$$
where $a \neq b$. We need to aim for an equation with N^2 on one side, so that we can use the method of the first part. We have $ab - \frac{1}{2}(a+b)N = 0$ which we write as
$$\left(a - \frac{N}{2}\right)\left(b - \frac{N}{2}\right) = \frac{N^2}{4} \quad \text{i.e.} \quad (2a - N)(2b - N) = N^2 .$$

Thus $2a - N = N^2$ and $2b - N = 1$ (or the other way round). The decomposition is therefore unique and given by
$$\frac{2}{N} = \frac{1}{\frac{1}{2}(N^2 + N)} + \frac{1}{\frac{1}{2}(N+1)} .$$

The only possible fly in the ointment is the $\frac{1}{2}$ in the denominators: a and b are supposed to be integers. However, all prime numbers greater than 2 are odd, so $N+1$ and $N^2 + N$ are both even and the denominators are indeed integers.

Post-mortem

Another way of getting the last part (less systematically) would have been to notice that
$$\frac{1}{N} = \frac{1}{(N+1)} + \frac{1}{N(N+1)} \implies \frac{2}{N} = \frac{1}{\frac{1}{2}(N+1)} + \frac{1}{\frac{1}{2}N(N+1)}$$
which gives the result immediately. How might you have noticed this? Well, I noticed it by trying to work out some examples, starting with what is given at the very beginning of the question. If $\dfrac{1}{3} = \dfrac{1}{4} + \dfrac{1}{12}$ then $\dfrac{2}{3} = \dfrac{2}{4} + \dfrac{2}{12}$ which works.

Of course, the advantage of the systematic approach is that it allows generalisation: what happens if N is odd but not prime; what happens if the numerator is 3 not 2?

Problem 12: Maximising with constraints (✓)

> Prove that the rectangle of greatest perimeter which can be inscribed in a given circle is a square. The result changes if, instead of maximising the sum of lengths of sides of the rectangle, we seek to maximise the sum of nth powers of the lengths of those sides for $n \geqslant 2$. What happens if $n = 2$? What happens if $n = 3$? Justify your answers.

1998 Paper I

Comments

Obviously, the perimeter of a *general* rectangle has no maximum (it can be as long as you like), so the key to this question is to use the constraint that the rectangle lies in the circle.

The vertices of a rectangle of greatest perimeter must lie *on* the circle; this seems obvious, but you should devote the first sentence of your solution to justifying it. And draw a diagram (I didn't because there wasn't enough room on the page).

The word 'greatest' in the question immediately suggests differentiating something. The perimeter of a rectangle is expressed in terms of two variables, length x and breadth y, so you must find a way of using the constraint to eliminate one variable. This could be done in two ways: express y in terms of x, or express both x and y in terms of another variable (an angle, say).

Finally, there is the matter of deciding whether your solution is the greatest or least value (or neither). This can be done either by considering second derivatives, or by trying to understand the different situations. Often the latter method, or a combination, is preferable.

The theory of stationary values of functions subject to constraints is very important: it has widespread applications (for example, in theoretical physics, financial mathematics and in fact in almost any area to which mathematics is applied). It forms a whole branch of mathematics called optimisation. Normally, the methods described above (eliminating one variable) are not used: a clever idea, the method of Lagrange multipliers, is used instead,

Solution to problem 12

The vertices of the rectangle must lie on the circle. Suppose not. Then two adjacent vertices (at least) lie in the interior of the circle. We could then increase the perimeter by extending two sides beyond these vertices. [Diagram here!]

Let the circle have diameter d and let the length of one side of the rectangle be x and the length of the adjacent side be y.

Then, by Pythagoras's theorem,
$$y = \sqrt{d^2 - x^2} \tag{\dag}$$
and the perimeter P is given by
$$P = 2x + 2\sqrt{d^2 - x^2}\,.$$

We can find the largest possible value of P as x varies by calculus. We have
$$\frac{\mathrm{d}P}{\mathrm{d}x} = 2 - 2\frac{x}{\sqrt{d^2 - x^2}}\,,$$
so for a stationary point we require (cancelling the factor of 2 and squaring)
$$1 = \frac{x^2}{d^2 - x^2} \quad \text{i.e.} \quad 2x^2 = d^2\,.$$

Thus $x = d/\sqrt{2}$ (ignoring $x = -d/\sqrt{2}$ for obvious reasons). Substituting this into (†) gives $y = d/\sqrt{2}$, so the rectangle is indeed a square, with perimeter $2\sqrt{d}$.

But is this the maximum perimeter? The easiest way to investigate is to calculate the second derivative, which is easily seen to be negative for all values of x, and in particular when $x = d/\sqrt{2}$. The stationary point is certainly a maximum. Alternatively, we could argue that the rectangle for which $x = 0$ has perimeter $2d$ which is less than $2\sqrt{2}d$ so $x = d/\sqrt{2}$ cannot correspond to the maximum perimeter.

For the second part, we consider
$$\mathrm{f}(x) = x^n + (d^2 - x^2)^{\frac{n}{2}}\,.$$

The first thing to notice is that f is constant if $n = 2$, so in this case, the largest (and smallest) value is d^2. You can use Pythagoras's theorem to see why this result holds.

For $n = 3$, we have
$$\mathrm{f}'(x) = 3x^2 - 3x(d^2 - x^2)^{\frac{1}{2}}\,,$$
so $\mathrm{f}(x)$ is stationary when $x^4 = x^2(d^2 - x^2)$, i.e. when $2x^2 = d^2$ as before or when $x = 0$. The corresponding stationary values of f are $\sqrt{2}d^3$ and $2d^3$, so this time the largest value occurs when $x = 0$.

Post-mortem

You will almost certainly want to investigate the situation for other values of n yourself, just to see what happens.

Were you happy with the proof given above that the square has the *largest* perimeter? And isn't it a bit odd that we did not discover a stationary point corresponding to the smallest value (which can easily be seen to occur at $x = 0$ or $x = d$)? To convince yourself that the maximum point gives the largest value, it is a good idea to sketch a graph: you can check that $\mathrm{d}P/\mathrm{d}x$ is positive for $0 < x < d/\sqrt{2}$ and negative for $d/\sqrt{2} < x < d$. The smallest values of the perimeter occur at the endpoints of the interval $0 \leqslant x \leqslant d$ and therefore do not have to be turning points.

A much better way of tacking the problem is to set $x = d\cos\theta$ and $y = d\sin\theta$ and find the perimeter as a function of θ. Because θ can take any value (there are no end points such as $x = 0$ to consider), the largest and smallest perimeters correspond to stationary points. Try it.

Problem 13: Binomial expansion (✓)

(i) Use the first four terms of the binomial expansion of $(1 - \frac{1}{50})^{\frac{1}{2}}$ to derive the approximation $\sqrt{2} \approx 1.414214$.

(ii) Calculate similarly an approximation to the cube root of 2 to six decimal places by considering $(1 + \frac{N}{125})^{\frac{1}{3}}$, where N is a suitable number.

[You need not justify the accuracy of your approximations.]

1998 Paper II

Comments

Although you do not have to justify your approximations, you do need to think carefully about the number of terms required in the expansions. You first have to decide how many you will need to obtain the given number of decimal places; then you have to think about whether the next term is likely to affect the value of the last decimal.

It is not at all obvious where the $\sqrt{2}$ in part (i) comes from until you write $(1 - \frac{1}{50})^{\frac{1}{2}} = \sqrt{\frac{49}{50}}$.

For part (ii), you have to choose N in such a way that $125 + N$ has something to do with a power of 2.

You can make the arithmetic of the binomial expansions a bit easier by arranging the denominator to be a power of 10 so that, for example, the expansion in part (i) becomes $(1 - \frac{2}{100})^{\frac{1}{2}}$. No need to use a calculator.

Solution to problem 13

(i) First we expand binomially:

$$\left(1 - \frac{2}{100}\right)^{\frac{1}{2}}$$

$$= 1 + \left(\frac{1}{2}\right)\left(-\frac{2}{100}\right) + \left(\frac{1}{2!}\right)\left(\frac{1}{2}\right)\left(-\frac{1}{2}\right)\left(-\frac{2}{100}\right)^2 - \left(\frac{1}{3!}\right)\left(\frac{1}{2}\right)\left(-\frac{1}{2}\right)\left(-\frac{3}{2}\right)\left(-\frac{2}{100}\right)^3 + \cdots$$

$$\approx 1 - \frac{1}{100} - \frac{5}{10^5} - \frac{5}{10^7} = 0.9899495\,.$$

It is clear that the next term in the expansion would introduce the eighth places of decimals, which it seems we do not need. Of course, after further manipulations we might find that the above calculation does not supply the 6 decimal places we need for $\sqrt{2}$, in which case we will have to work out the next term in the expansion.

But $\left(\dfrac{98}{100}\right)^{\frac{1}{2}} = \dfrac{7\sqrt{2}}{10}$, so $\sqrt{2} \approx 9.899495/7 \approx 1.414214$.

(ii) We have

$$\left(1 + \frac{3}{125}\right)^{\frac{1}{3}}$$

$$= 1 + \left(\frac{1}{3}\right)\left(\frac{3}{125}\right) + \left(\frac{1}{2!}\right)\left(\frac{1}{3}\right)\left(-\frac{2}{3}\right)\left(\frac{3}{125}\right)^2 + \left(\frac{1}{3!}\right)\left(\frac{1}{3}\right)\left(-\frac{2}{3}\right)\left(-\frac{5}{3}\right)\left(\frac{3}{125}\right)^3 + \cdots$$

$$\approx 1 + \frac{8}{1000} - \frac{64}{10^6} + \frac{5}{3}\frac{8^3}{10^9}$$

$$= 1.007936 + \frac{256}{3}\frac{1}{10^8} = 1.007937\,.$$

Successive terms in the expansion decrease by a factor of about 1000, so this should give the right number of decimal places.

But $\left(\dfrac{128}{125}\right)^{\frac{1}{3}} = \dfrac{4\sqrt[3]{2}}{5} = \dfrac{8\sqrt[3]{2}}{10}$ so $\sqrt[3]{2} \approx \dfrac{10.0793}{8} \approx 1.259921$.

Post-mortem

This question required a bit of intuition, and some accurate arithmetic. You don't have to be brilliant at arithmetic to be a good mathematician, but most mathematicians aren't bad at it. There have in the past been children who were able to perform extraordinary feats of arithmetic. For example, Zerah Colburn, a 19th century American, toured Europe at the age of 8. He was able to multiply instantly any two four digit numbers given to him by the audiences. George Parker Bidder (the Calculating Boy) could perform similar feats, though unlike Colburn he became a distinguished mathematician and scientist. One of his brothers knew the bible by heart.

The error in the approximation is the weak point of this question. Although it is clear that the next term in the expansion is too small to affect the accuracy, it is not obvious that the sum of all the next hundred (say) terms of the expansion is negligible (though in fact it is). What is needed is an estimate of the truncation error in the binomial expansion. Such an estimate is not hard to obtain (first year university work) and is typically of the same order of magnitude as the first neglected term in the expansion. Without this estimate, the approximation is not justified.

Problem 14: Sketching subsets of the plane (✓✓)

Sketch the following subsets of the x-y plane:

(i) $|x| + |y| \leqslant 1$;

(ii) $|x-1| + |y-1| \leqslant 1$;

(iii) $|x-1| - |y+1| \leqslant 1$;

(iv) $|x|\,|y-2| \leqslant 1$.

1999 Paper I

Comments

Often with modulus signs, it is easiest to consider the cases separately, so for example in part (i), you would first work out the case $x > 0$ and $y > 0$, then $x > 0$ and $y < 0$, and so on. Here there is a simple geometric understanding of the different cases: once you have worked out the first case, the other three can be deduced by symmetry.

Another geometric idea should be in your mind when tackling this question, namely the idea of translations in the plane.

Post-mortem

There was no room for a post-mortem over the page, because the diagrams take up so much space.

Don't read this until you have tried the question!

There are two key learning points, if you will excuse this horrible expression.

The first is that if you want to sketch a region, it is often best to draw the curves that define the boundary of the region, then just work out whether you want the interior or exterior of the boundary by choosing one interior point and seeing whether it satisfies the inequalities.

The second is that quite complicated inequalities can sometimes be much simplified by translating or rotating the axes.

Solution to problem 14

The way to deal with the modulus signs in this question is to consider first the case when the things inside the modulus signs are positive, and then get the full picture by symmetry, shifting the origin as appropriate.

For part (i), consider the first quadrant $x \geqslant 0$ and $y \geqslant 0$. In this quadrant, the inequality is $x+y \leqslant 1$. Draw the line $x+y=1$ and then decide which side of the line is described by the inequality. It is obviously (since x has to be smaller than something) the region to the left of the line; or (since y also has to be smaller than something) the region below the line, which is the same region.

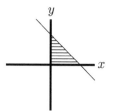

Similar arguments could be used in the other quadrants, but it is easier to note that the inequality $|x|+|y| \leqslant 1$ is unchanged when x is replaced by $-x$, or y is replaced by $-y$, so the sketch should have reflection symmetry in both axes, as shown in the diagram on the right.

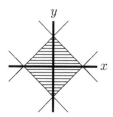

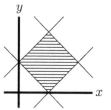

Part (ii) is the same as part (i), except for a translation. The origin of (i) is now at $(1,1)$.

For part (iii), consider first $x - y \leqslant 1$, instead of $|x-1| - |y+1| \leqslant 1$. For $x > 0$ and $y > 0$, this gives a region that is infinite in extent in the positive y direction, as shown in the first diagram below. Then reflect this in both axes and translate one unit down the y axis and 1 unit along the positive x axis as shown in the second and third diagrams below.

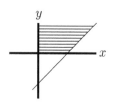 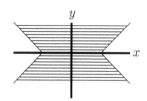 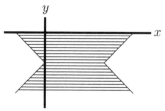

For part (iv), consider first the rectangular hyperbola $xy = 1$ reflected in both axes to give four hyperbolas, as in the first diagram below. Then translate the four hyperbolas translated 2 units up the y axis as in the final diagram. The region required is enclosed by the four hyperbolas.

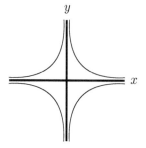 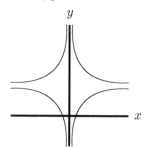

Problem 15: More sketching subsets of the plane (✓✓)

(i) Show that
$$x^2 - y^2 + x + 3y - 2 = (x - y + 2)(x + y - 1)$$
and hence, or otherwise, indicate by means of a sketch the region of the x-y plane for which
$$x^2 - y^2 + x + 3y > 2.$$

(ii) Sketch also the region of the x-y plane for which
$$x^2 - 4y^2 + 3x - 2y < -2.$$

(iii) Give the coordinates of a point for which both inequalities are satisfied or explain why no such point exists.

1995 Paper I

Comments

This question gets a ✓✓ difficulty rating because inequalities always need to be handled with care.

For the very first part, you could either factorise the left hand side to obtain the right hand side, or multiply out the right hand side to get the left hand side. Obviously, it is much easier to do the multiplication than the factorisation. But is that 'cheating' or taking a short cut that might lose marks? No, it is not cheating: it doesn't matter if you start from the given answer and work backwards — it is still a mathematical proof and any proof will get the marks. (But note that if there is a 'hence' in the question, you will lose marks if you do not use the result or results that you have just proved.)

In part (ii), you have to do the factorisation yourself, so you should look carefully at where the terms in the (very similar) first part came from. It will help to spot the similarities between $x^2 - y^2 + x + 3y - 2$ and $x^2 - 4y^2 + 3x - 2y + 2$.

Since no indication is given as to what detail should appear on the sketch, you have to use your judgement: it is clearly important to know where the regions lie relative to the coordinate axes.

Post-mortem

As in the previous sketching question, the solution is so long that there is no room for a post-mortem after the answer.

Don't read this before having tried the question!

The answer given overleaf is inadequate in two ways, each of which would have probably lost me marks. First, as in the previous question, the answer should definitely have referred to the boundaries. Here, the inequalities are *strict*, which means that the boundary lines are *excluded* from the required regions. Second, justification should have been provided for the claim that the point $(1, 2)$ satisfies both inequalities. This could be provided either by simply indicating the position of the point on both sketches or algebraically by substitution into the inequalities.

Solution to problem 15

To do the multiplication, it pays to be systematic and to set out the algebra nicely:

$$\begin{aligned}(x - y + 2)(x + y - 1) &= x(x + y - 1) - y(x + y - 1) + 2(x + y - 1) \\ &= x^2 + xy - x - yx - y^2 + y + 2x + 2y - 2 \\ &= x^2 + x - y^2 + 3y - 2\end{aligned}$$

as required.

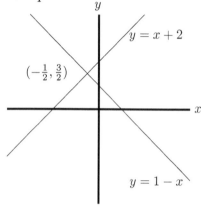

For the first inequality, we need both $x - y + 2$ and $x + y - 1$ either to be positive or to be negative. To sort out the inequalities (or inequations as they are sometimes horribly called), the first thing to do is to draw the lines corresponding to the corresponding equalities. These lines divide the plane into four regions and we then have to decide which regions are relevant.

The diagonal lines $x - y = -2$ and $x + y = 1$ intersect at $(-\frac{1}{2}, \frac{3}{2})$; this is the important point to mark in on the sketch. The required regions are the left and right quadrants formed by these diagonal lines (since the inequalities mean that the regions are to the right of both lines or to the left of both lines).

For part (ii), the first thing to do is to factorise $x^2 - 4y^2 + 3x - 2y + 2$. The key similarity with the first part is the absence of 'cross' terms of the form xy. This allows a difference-of-two-squares factorisation of the first two terms: $x^2 - 4y^2 = (x - 2y)(x + 2y)$. Following the pattern of the first part, we can then try a factorisation of the form

$$x^2 - 4y^2 + 3x - 2y + 2 = (x - 2y + a)(x + 2y + b)$$

where $ab = 2$. Considering the terms linear in x and y gives $a + b = 3$ and $2a - 2b = -2$ which quickly leads to $a = 1$ and $b = 2$. Note that altogether there were three equations for a and b, so we had no right to expect a consistent solution (except for the fact that this is a STEP question for which we had every right to believe that the first part would guide us through the second part).

Alternatively, we could have completed the square in x and in y and then used difference of two squares:

$$\begin{aligned}x^2 - 4y^2 + 3x - 2y + 2 &= (x + \tfrac{3}{2})^2 - \tfrac{9}{4} - (2y + \tfrac{1}{2})^2 + \tfrac{1}{4} + 2 \\ &= (x + \tfrac{3}{2})^2 - (2y + \tfrac{1}{2})^2 = (x + \tfrac{3}{2} - 2y - \tfrac{1}{2})(x + \tfrac{3}{2} + 2y + \tfrac{1}{2}).\end{aligned}$$

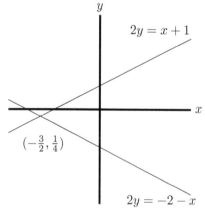

As in the previous case, the required region is formed by two intersecting lines; this time, they intersect at $(-\frac{3}{2}, -\frac{1}{4})$ and the upper and lower regions are required, since the inequality is the other way round.

It is easy to see from the sketches that there are points that satisfy both inequalities: for example $(1, 2)$.

Problem 16: Non-linear simultaneous equations (✓)

Consider the system of equations

$$2yz + zx - 5xy = 2$$
$$yz - zx + 2xy = 1$$
$$yz - 2zx + 6xy = 3.$$

Show that

$$xyz = \pm 6$$

and find the possible values of x, y and z.

1996 Paper II

Comments

At first sight, this looks forbidding. A closer look reveals that the variables x, y and z occur only in pairs yz, zx and xy. The problem therefore boils down to solving three simultaneous equations in the variables yz, zx and xy, then using the solution to find x, y and z individually.

What do you make of the equation $xyz = \pm 6$? How could the $\pm$ arise?

There are two ways of tackling such simultaneous equations. You could use the first equation to find an expression for one variable (yz say) in terms of the other two variables, then substitute this into the other equations to eliminate yz from the system. Then use the second equation (in its new form) to find an expression for one of the two remaining variables (zx say), and substitute this into the third equation (in its new form) to obtain an equation for the third variable (xy). Having solved this equation, you can substitute back to find the other variables. This method is called *Gauss elimination*.

Alternatively, you could eliminate one variable (yz say) from the first two of equations by multiplying the first equation by something suitable and the second equation by something suitable and subtracting. You then eliminate yz from the second and third equations similarly. That leaves you with two simultaneous equations in two variables which you can solve by your favourite method.

There is another way of solving the simultaneous equations, which is better in theory than in practice. You write the equations in matrix form $\mathbf{Mx} = \mathbf{c}$, where in this case

$$\mathbf{M} = \begin{pmatrix} 2 & 1 & -5 \\ 1 & -1 & 2 \\ 1 & -2 & 6 \end{pmatrix}, \quad \mathbf{x} = \begin{pmatrix} yz \\ zx \\ xy \end{pmatrix}, \quad \mathbf{c} = \begin{pmatrix} 2 \\ 1 \\ 3 \end{pmatrix}.$$

The solution is then $\mathbf{x} = \mathbf{M}^{-1}\mathbf{c}$. All you have to do is invert a 3×3 matrix, which is doable but not very pleasant. You might like to try it if you know the formula for the inverse of a matrix.

Solution to problem 16

Start by labelling the equations:
$$2yz + zx - 5xy = 2, \tag{1}$$
$$yz - zx + 2xy = 1, \tag{2}$$
$$yz - 2zx + 6xy = 3. \tag{3}$$

We use Gaussian elimination. Rearranging equation (1) gives
$$yz = -\tfrac{1}{2}zx + \tfrac{5}{2}xy + 1, \tag{4}$$

which we substitute back into equations (2) and (3):
$$-\tfrac{3}{2}zx + \tfrac{9}{2}xy = 0, \tag{5}$$
$$-\tfrac{5}{2}zx + \tfrac{17}{2}xy = 2. \tag{6}$$

Thus $zx = 3xy$ (using equation (5)). Substituting into equation (6) gives $xy = 2$ and $zx = 6$. Finally, substituting back into equation (1) shows that $yz = 3$.

The question is now plain sailing. Multiplying the three values together gives $(xyz)^2 = 36$ and taking the square root gives $xyz = \pm 6$ as required.

Now it remains to solve for x, y and z individually. We know that $yz = 3$, so if $xyz = +6$ then $x = +2$, and if $xyz = -6$ then $x = -2$. The solutions are therefore either $x = +2$, $y = 1$, and $z = 3$ or $x = -2$, $y = -1$, and $z = -3$.

Post-mortem

There were two key observations which allowed us to do this question quite easily. Both came from looking carefully at the question. The first was that the given equations, although non-linear in x, y and z (they are quadratic, since they involve products of these variables) could be thought of as three linear equations in yz, zx and xy. That allowed us to make a start on the question. The second observation was that the equation $xyz = \pm 6$ is almost certain to come from $(xyz)^2 = 36$ and that gave us the next step after solving the simultaneous equations. (Recall the next step was to multiply all the variables together.)

There was a point of technique in the solution: it is often very helpful in this sort of problem (and many others) to number your equations. This allows you to refer back clearly and quickly, for your benefit as well as for the benefit of your readers.

The three simultaneous quadratic equations (1) – (3) have a geometric interpretation but it is not at all obvious. The equations are quadratic in the variables x, y and z, which means that each equation represents either an ellipsoid or a hyperboloid (or some special cases).[17] The solutions of all three equations represent the points of intersection of the three surfaces. Not very easy to picture.

---

[17] An *ellipsoid* is roughly the shape of the surface of a rugby ball, or of the giant galaxy ESO 325-G004. A *hyperboloid* can either be the shape of an infinite radar dish (in fact, a pair of such dishes) or it can be the shape of a power station cooling tower. Our equations in fact represent hyperboloids of the cooling tower type.

Problem 17: Inequalities (✓✓)

> Solve the inequalities
>
> (i) $1 + 2x - x^2 > \dfrac{2}{x}$ $(x \neq 0)$,
>
> (ii) $\sqrt{3x+10} > 2 + \sqrt{x+4}$ $(x \geq -10/3)$.

2001 Paper I

Comments

The two parts are unrelated (unusually for STEP questions), except that they deal with inequalities. Both parts need care, having traps for the unwary.

In part (i) you have to watch out when you multiply an inequality: if the thing you multiply by is negative then the inequality reverses. You might find sketching a graph helpful.

In part (ii) you have to consider the possibility that your algebraic manipulations have created extra spurious solutions. It is worth (after you have finished the question) sketching the graphs of $\sqrt{3x+10}$ and $2 + \sqrt{x+4}$ just to see what is going on. You can get the latter graph by translations of $\sqrt{x}$ (and you can get $y = \sqrt{x}$ by reflection $y = x^2$ in the line $y = x$). I would have drawn them for you in the post-mortem overleaf had there been room.

In fact, a solution relying on sketches for part (ii) is probably preferable to my solution. With the sketches to hand, you only have to solve the equation $\sqrt{3x+10} = 2 + \sqrt{x+4}$ and then look at your sketches to see to see what range of values of x you need, thereby saving much anguish.[18]

---

[18] Now I look at it again, it seems to me that the last sentence of my solution ('Therefore the inequality holds ...') is a bit suspect without sketches to show that the inequality actually *does* hold.

Solution to problem 17

(i) We would like to multiply both sides of the inequality by x in order to obtain a nice cubic expression. However, we have to allow for the possibility that x is negative (which would reverse the inequality). One way of dealing with this is to consider the cases $x > 0$ and $x < 0$ separately.

For $x > 0$, we multiply the whole equation by x without changing the direction of the inequality:

$$1 + 2x - x^2 > \frac{2}{x} \Rightarrow x + 2x^2 - x^3 > 2$$
$$\Rightarrow x^3 - 2x^2 - x + 2 < 0$$
$$\Rightarrow (x-1)(x+1)(x-2) < 0$$
$$\Rightarrow 1 < x < 2, \tag{*}$$

discarding the possibility $x < -1$, since we have assumed that $x > 0$.

The easiest way of obtaining the result $(*)$ from the previous line is to sketch the graph of $(x-1)(x+1)(x-2) < 0$.
For $x < 0$, we must reverse the inequality when we multiply by x so in this case, $(x-1)(x+1)(x-2) > 0$, which gives $-1 < x < 0$.
In the sketch on the right, the values of x for which the inequality holds are shown as bold lines.

The smart way to do $x > 0$ and $x < 0$ in one step is to multiply by x^2 (which is never negative and hence never changes the direction of the inequality) and analyse $x(x-1)(x+1)(x-2) < 0$. Again, a sketch is useful.

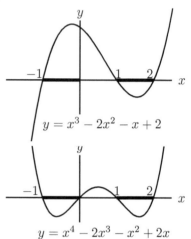

(ii) First square both sides the inequality $\sqrt{3x+10} > 2 + \sqrt{x+4}$:

$$3x + 10 > 4 + 4\sqrt{x+4} + (x+4) \quad \text{i.e.} \quad x + 1 > 2\sqrt{x+4}\,.$$

Note that the both sides of the original inequality are positive or zero (i.e. non-negative), so the direction of the inequality is not changed by squaring. Now consider the new inequality $x + 1 > 2\sqrt{x+4}$. If both sides are non-negative, that is if $x > -1$, we can square both sides again without changing the direction of the inequality. But, if $x < -1$, the inequality cannot be satisfied since the right hand side is always non-negative. Squaring gives

$$x^2 + 2x + 1 > 4(x+4) \quad \text{i.e.} \quad (x-5)(x+3) > 0\,.$$

Thus, $x > 5$ or $x < -3$. However, we must reject $x < -3$ because of the condition $x > -1$. Therefore the inequality holds for $x > 5$.

Post-mortem

The spurious result $x < -3$ at the end of part (ii) arises through loss of information in the process of squaring: if you square an expression, you lose its sign. After squaring twice, the resulting inequality is the same as would have resulted from $\sqrt{(3x+10)} < 2 - \sqrt{(x+4)}$ and it is to this inequality that $-\frac{10}{3} \leqslant x < -3$ is a solution.

Problem 18: Inequalities from cubics (✓✓)

(i) Sketch, without calculating the stationary points, the graph of the function $f(x)$ given by

$$f(x) = (x-p)(x-q)(x-r),$$

where $p < q < r$. By considering the quadratic equation $f'(x) = 0$, or otherwise, show that

$$(p+q+r)^2 > 3(qr+rp+pq).$$

(ii) By considering $(x^2+gx+h)(x-k)$, or otherwise, show that $g^2 > 4h$ is a sufficient condition but not a necessary condition for the inequality

$$(g-k)^2 > 3(h-gk)$$

to hold.

2001 Paper

Comments

The idea behind the first part is to obtain an inequality by considering a certain graph. In the second part, we again use graphs to obtain an inequality, which this time holds subject to a different inequality.

For the sketch in the first part it is not necessary to do more than think about the behaviour for x large and positive, and for x large and negative, and the points at which the graph crosses the x-axis.

For the second part, you should also have a sketch or sketches in mind. The argument is very similar to that of the first part, except that it also tests understanding of the meaning of the terms necessary and sufficient. For this, it is probably best to use $\Rightarrow$ notation.

To show that $g^2 > 4h$ is not a *necessary* condition for the inequality $(g-k)^2 > 3(h-gk)$ to hold, you just have to give an example (the simpler the better) for which $(g-k)^2 > 3(h-gk)$ but $g^2 \leqslant 4h$.

Solution to problem 18

From the sketch, we see that $f(x)$ has two turning points, so the equation $f'(x) = 0$ has two real roots.
Now

$$f(x) = (x-p)(x-q)(x-r)$$
$$= x^3 - (p+q+r)x^2 + (qr+rp+pq)x - pqr$$

so at a turning point

$$f'(x) = 3x^2 - 2(p+q+r)x + (qr+rp+pq) = 0 .$$

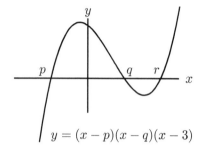

$y = (x-p)(x-q)(x-3)$

Using the condition '$b^2 > 4ac$' for this quadratic to have two real root gives the required result:

$$4(p+q+r)^2 > 12(qr+rp+pq) .$$

For the second part, we can use a similar argument. First note that if the quadratic equation $x^2 + gx + h = 0$ has two distinct real roots then the cubic equation $(x^2 + gx + h)(x - k) = 0$ has three roots, at least two of which are distinct. Thus

$$g^2 > 4h \Rightarrow (x^2 + gx + h)(x - k) = 0 \text{ has at least two distinct roots}$$
$$\Rightarrow (x^2 + gx + h)(x - k) \text{ has at least two distinct turning points (draw graphs!)}$$
$$\Rightarrow x^3 + (g - k)x^2 + (h - gk)x - hk \text{ has at least two distinct turning points}$$
$$\Rightarrow 3x^2 + 2(g - k)x + (h - gk) = 0 \text{ has two distinct roots}$$
$$\Rightarrow 4(g - k)^2 > 12(h - gk) \text{ as required.}$$

Thus $g^2 > 4h$ is a sufficient condition for this inequality to hold.

Why is it not necessary? That is to say, why is it too strong a condition? The reason is that both turning points could be above the x–axis or both could be below (not one on either side which was the origin of the inequality). Saying this would get all the marks. Or you could give a counterexample: for example, $g = 2, h = 1$ and $k = 1000$.

Post-mortem

When I set this question, I got considerable satisfaction from the way that seemingly obscure inequalities are derived from understanding simple graphs and the quadratic formula. It was only when preparing this book that I realised that they are, disappointingly, not at all obscure.

The inequality of the first part is equivalent to $(q - r)^2 + (r - p)^2 + (p - q)^2 > 0$, the inequality coming from the fact the squares of real numbers are non-negative and in this case the given condition $p < q < r$ means that the expression is strictly positive (not equal to zero).

If we rewrite the second inequality as $(k + \tfrac{1}{2}g)^2 + 3(\tfrac{1}{4}g^2 - h) > 0$ we see that it certainly holds if $(\tfrac{1}{4}g^2 - h) > 0$, as expected. This is not a necessary condition: it also holds if $(\tfrac{1}{4}g^2 - h) < 0$ provided k is large enough.

Problem 19: Logarithms (✓✓✓)

To nine decimal places, $\log_{10} 2 = 0.301029996$ and $\log_{10} 3 = 0.477121255$.

(i) Calculate $\log_{10} 5$ and $\log_{10} 6$ to three decimal places. By taking logs, or otherwise, show that
$$5 \times 10^{47} < 3^{100} < 6 \times 10^{47}.$$
Hence write down the first digit of 3^{100}.

(ii) Find the first digit of each of the following numbers: 2^{1000}; $2^{10\,000}$; and $2^{100\,000}$.

2000 Paper I

Comments

This nice little question shows why it is a good idea to ban calculators from some mathematics examinations — though I notice that my calculator can't work out 2^{10000}.

When I was at school, before electronic calculators were invented, we had to spend quite a lot of time in year 8 (I think) doing extremely tedious calculations by logarithms. We were provided with a book of tables of four-figure logarithms and lots of uninteresting numbers to multiply or divide. The book also had tables of trigonometric functions. When it came to antilogging, to get the answer, we had to use the tables backwards, since no tables of inverse logarithms (exponentials) were provided.

A logarithm (to base 10, as always in this question) consists of two parts: the *characteristic* which is the number before the decimal point and the *mantissa* which is the number after the decimal point. It is the mantissa that gives the significant figures of the number that has been logged; the characteristic tells you where to put the decimal point. The important property of logs to base 10 is that $A \times 10^n$ and $A \times 10^m$ have the same mantissa, so log tables need only show the mantissa.

The characteristic of a number greater than 1 is non-negative but the characteristic of a number less than 1 is negative. The rules for what to do in the case of a negative characteristic were rather complicated: you couldn't do ordinary arithmetic because the logarithm consisted of a negative characteristic and a positive mantissa. In ordinary arithmetic, the number -3.4 means $-3 - 0.4$ whereas the corresponding situation in logarithms, normally written $\bar{3}.4$, means $-3 + 0.4$. Instead of explaining this, the teacher gave a complicated set of rules, which just had to be learned — not the right way to do mathematics.

This question is all about calculating mantissas and there are no negative characteristics, I'm happy to say.

Solution to problem 19

(i) For the very first part, we have

$\log 2 + \log 5 = \log 10 = 1$, so $\log 5 = 1 - 0.301029996 = 0.699$ (3 d.p.)

and

$\log 6 = \log 2 + \log 3 = 0.778$ (3 d.p.).

Now we have to show that

$$5 \times 10^{47} < 3^{100} < 6 \times 10^{47}.\quad (*)$$

Taking logs preserves the inequalities (because $\log x$ is an *increasing* function), so we need to show that

$$47 + \log 5 < 100 \log 3 < 47 + \log 6$$

i.e. that

$$47 + 0.699 < 47.7121 < 47 + 0.778$$

which is true. We see from $(*)$ that the first digit of 3^{100} is 5.

(ii) To find the first digit of these numbers, we use the method of part (i).

We have (to 3 d.p.)

$\log 2^{1000} = 1000 \log 2 = 301.030 = 301 + 0.030 < 301 + \log 2$

Thus $10^{301} < 2^{1000} < 10^{301} \times 2$ and the first digit of $2^{1\,000}$ is 1.

Similarly,

$\log 2^{10\,000} = 10\,000 \log 2 = 3010.29996 < 3010 + \log 2$, so the first digit of $2^{10\,000}$ is 1.

Finally,

$\log 2^{100\,000} = 30102 + 0.9996$ (4 d.p.) and

$\log 9 = 2 \log 3 = 0.95$ (2 d.p.), so the first digit of $2^{100\,000}$ is 9.

Post-mortem

Although the ideas in this question are really quite elementary, you needed to understand them deeply. You should feel pleased with yourself it you got this one out.

Note that in part (i) I started with $(*)$, the result I was trying to prove and then showed it is true. This is of course dangerous. But provided you keep writing 'We have to prove that ... ' or 'RTP' (Required To Prove) you should not get muddled between what you have proved and what you are trying to prove.

Problem 20: Cosmological models (✓✓)

In a cosmological model, the radius R of the universe is a function of the age t of the universe. The function R satisfies the three conditions:

$$R(0) = 0, \qquad R'(t) > 0 \text{ for } t > 0, \qquad R''(t) < 0 \text{ for } t > 0, \qquad (*)$$

where R″ denotes the second derivative of R. The function H is defined by

$$H(t) = \frac{R'(t)}{R(t)}.$$

(i) Sketch a graph of $R(t)$. By considering a tangent to the graph, show that $t < \dfrac{1}{H(t)}$.

(ii) Observations reveal that $H(t) = \dfrac{a}{t}$, where a is constant. Derive an expression for $R(t)$.

What range of values of a is consistent with the three conditions $(*)$?

(iii) Suppose, instead, that observations reveal that $H(t) = bt^{-2}$, where b is constant. Show that this is not consistent with conditions $(*)$ for any value of b.

Note: $x^\alpha e^{-x} \to 0$ as $x \to \infty$ for any constant α.

2001 Paper I

Comments

The sketch just means any graph starting at the origin and increasing, but with decreasing gradient. The second part of (i) needs a bit of thought (where does the tangent intersect the x-axis?) so don't despair if you don't see it immediately. Parts (ii) and (iii) are perhaps easier than part (i).

Solution to problem 20

(i) The figure shows the curve $y = R(t)$ and the tangent to the curve, which meets the t axis.
The height of the right-angled triangle in the figure is $R(t)$ and the slope of the hypotenuse is $R'(t)$. The length of the base is therefore $R(t)/R'(t)$, i.e. $1/H(t)$.
The figure shows that the tangent to the graph for $t > 0$ intersects the *negative* t–axis, so $1/H(t) > t$.

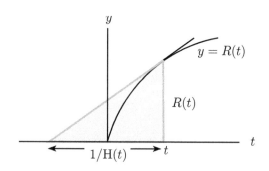

(ii) One way to proceed is to integrate the differential equation:

$$H(t) = \frac{a}{t} \Rightarrow \frac{R'(t)}{R(t)} = \frac{a}{t} \Rightarrow \int \frac{R'(t)}{R(t)} dt = \int \frac{a}{t} dt \Rightarrow \ln R(t) = a \ln t + \text{constant} \Rightarrow R(t) = At^a.$$

The first two conditions ($*$) are satisfied if $a > 0$ and $A > 0$. For the third condition, we have $R''(t) = a(a-1)At^{a-2}$ which is negative provided $a < 1$. The range of a is therefore $0 < a < 1$.

(iii) The obvious way to do this just follows part (ii). This time, we have

$$H(t) = \frac{b}{t^2} \Rightarrow \frac{R'(t)}{R(t)} = \frac{b}{t^2} \Rightarrow \int \frac{R'(t)}{R(t)} dt = \int \frac{b}{t^2} dt \Rightarrow \ln R(t) = -\frac{b}{t} + \text{constant} \Rightarrow R(t) = Ae^{-b/t}.$$

Thus $R'(t) = H(t)R(t) = Abt^{-2}e^{-b/t}$. Clearly $A > 0$ since $R(t) > 0$ for $t > 0$ so $R'(t) > 0$ provided $b > 0$. Furthermore $R(0) = 0$ (think about this!), so only the condition $R''(t) < 0$ remains to be checked. Differentiating $R'(t)$ gives

$$R''(t) = Ab(-2t^{-3})e^{-b/t} + Abt^{-2}e^{-b/t}(bt^{-2}) = Abt^{-4}e^{-b/t}(-2t + b)$$

This is positive when $t < \tfrac{1}{2}b$, which contradicts the condition $R'' < 0$.

Instead of solving the differential equation, we could proceed as follows. We have

$$\frac{R'}{R} = bt^{-2} \Rightarrow R' = bRt^{-2} \Rightarrow R'' = bR't^{-2} - 2bRt^{-3} = b^2t^{-4}R - 2bRt^{-3} = b(b-2t)t^{-4}R.$$

This is positive when $t < \tfrac{1}{2}b$, which contradicts the condition $R'' < 0$.

Post-mortem

The interpretation of the conditions in ($*$) of the question is as follows. The first condition $R(0) = 0$ says that the universe started from zero radius — in fact, from the 'big bang'.

The second condition says that the universe is expanding. This was one of the key discoveries in cosmology in the last century. The quantity $H(t)$ is called *Hubble's constant* (though it varies with time). It measures the rate of expansion of the universe. Its reciprocal is the Hubble time. The present-day value of the Hubble time has been a matter of great debate but it seems to be about 14.4 billion years, greater than the age of the universe (as calculated using cosmological models of this sort) by about 0.5 billion years.

The last condition says that the expansion of the universe is slowing down. This is expected on physical grounds because of the gravitational attraction of galaxies on one another. However, current observations indicate that the expansion of the universe is speeding up, and this is thought to be due to the presence of dark energy.

Problem 21: Melting snowballs (✓✓)

Frosty the snowman is made from two uniform spherical snowballs, initially of radii $2R$ and $3R$. The smaller (which is his head) stands on top of the larger. As each snowball melts, its volume decreases at a rate which is directly proportional to its surface area, the constant of proportionality being the same for each snowball. During melting, the snowballs remain spherical and uniform. When Frosty is half his initial height, show that the ratio of his volume to his initial volume is $37 : 224$.

What is this ratio when Frosty is one tenth of his initial height?

1991 Paper I

Comments

To start with, you have to set up a differential equation which gives the radii of the snowballs as a function of time. Don't worry if you have never solved a differential equation: you will be able to solve this one. Having solved it, you have to evaluate the constant of integration from the information in the question.

Setting up and solving the differential equation is not difficult in itself, but it is necessary to think what variables you want to use and then go through a number of steps in the dark, without any reassurance from the question.

Solution to problem 21

For either snowball, the rate of change of volume, call it v, at any time is related to the surface area a by

$$\frac{dv}{dt} = -ka, \qquad (*)$$

where k is a positive constant. For a sphere of radius r, this becomes

$$\frac{d}{dt}(\tfrac{4}{3}\pi r^3) = -k(4\pi r^2).$$

We can write $\dfrac{d(r^3)}{dt}$ as $3r^2 \dfrac{dr}{dt}$, so equation $(*)$ is equivalent to (cancelling the factor of $4\pi r^2$)

$$\frac{dr}{dt} = -k.$$

Thus $r = -kt + C$, where C is a constant of integration.

Initially, Frosty's head has radius $2R$ and his body has radius $3R$, so the equations for the radii of the head and body at time t are respectively

$$r = -kt + 2R \quad \text{and} \quad r = -kt + 3R.$$

Frosty's height h is twice the sum of these radii, i.e. $h = 2(-2kt + 5R)$, which falls to half its original value of $10R$ when $kt = \tfrac{5}{4}R$. At this time, the radii of the head and body are $\tfrac{3}{4}R$ and $\tfrac{7}{4}R$, so the ratio of his volume to his initial volume is

$$\frac{(\tfrac{4}{3}\pi)(\tfrac{3}{4}R)^3 + (\tfrac{4}{3}\pi)(\tfrac{7}{4}R)^3}{(\tfrac{4}{3}\pi)(2R)^3 + (\tfrac{4}{3}\pi)(3R)^3} = \frac{(\tfrac{3}{4})^3 + (\tfrac{7}{4})^3}{2^3 + 3^3} = \frac{37}{224}.$$

When Frosty is just R high, all that remains of him, since $kt > 2R$, is the body, which is a sphere of radius $\tfrac{1}{2}R$. The ratio of his volume to his original volume is just $\tfrac{1}{240}$.

Post-mortem

As mentioned in the comments section, there is not a lot for you to do in this question, but it is a lot for you to do entirely on your own.

The key is to transcribe the words in the question into mathematics, the obvious starting point being equation $(*)$.

The original question had a different rider[19], asking about the maximum rate of change of volume with respect to area. I didn't much like this, because it seemed to be a new and not very interesting idea. My rider relates directly to the previous part, and requires a tiny bit of extra thought (a trap, some would say, but it is only a trap if you are on autopilot).

---

[19] The term *rider* seems a bit old-fashioned now. It referred to the final part at the end of an examination question in the days when examination questions often consisted of a relatively straightforward first part, perhaps the proof of a theorem, followed by a more tricky part extending or applying the first part.

Problem 22: Gregory's series (✓✓)

Give rough sketches of the function $\tan^k \theta$ for $0 \leq \theta \leq \frac{1}{4}\pi$ in the two cases $k = 1$ and $k \gg 1$.

(i) Show that for any positive integer n

$$\int_0^{\frac{1}{4}\pi} \tan^{2n+1}\theta \, d\theta = (-1)^n \left(\tfrac{1}{2}\ln 2 + \sum_{m=1}^{n} \frac{(-1)^m}{2m} \right), \qquad (\dagger)$$

and deduce that

$$\ln 2 = -\sum_{m=1}^{\infty} \frac{(-1)^m}{m}. \qquad (\ddagger)$$

(ii) Show similarly that

$$\frac{\pi}{4} = -\sum_{m=1}^{\infty} \frac{(-1)^m}{2m-1}.$$

1991 Paper II

Comments

The symbol $\gg$ in the first paragraph means 'much greater than', so for the second sketch k is a large number.

This is a good question. In part (i) you are told what to do (in not-very-easy stages) and part (ii) tests your understanding of what you have done and why you have done it by asking you to apply the method to a different but essentially similar problem.

In the first paragraph, you have to see how the function $\tan^k \theta$ changes when k increases. You need only a rough sketch to show that you have understood the important point. This should be done by thought, not by means of a calculator.

If you are stuck with the integral of the second paragraph, you might like to think in terms of a recurrence formula, i.e. a formula relating I_{2n+1} and I_{2n-1} (in the obvious notation).

The series derived in part (ii) for $\frac{1}{4}\pi$ is usually called Leibniz' formula, although the general series for $\tan^{-1} x$ was written down by Gregory in 1671, two years before Leibniz. It was one of the first explicit formulae for π, though Wallis had obtained a product formula in 1655 using a method similar to the method of this question, using $\sin^k x$ in the integral. Previously, the value of π could only be estimated geometrically, by (for example) approximating the circumference of a circle by the edges of an inscribed regular polygon. Using a square gives $\pi \approx 2\sqrt{2}$.

Solution to problem 22

For $0 \leqslant \tan\theta < \frac{1}{4}\pi$, we have $\tan\theta < 1$, so that the curve $y = \tan^k \theta$ is close to zero (i.e. much smaller than 1) when k is large. This is illustrated in the figure which shows three cases: for $k = 1$ the graph is mildly curved; for larger k the graph hugs the x-axis before taking off. The graphs all pass through the point $(\frac{1}{4}\pi, 1)$.

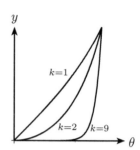

(i) To evaluate the integral, let

$$I_{2n+1} = \int_0^{\frac{1}{4}\pi} \tan^{2n+1}\theta \, d\theta. \qquad (**)$$

We shall express I_{2n+1} in terms of I_{2n-1} using the relation $\tan^2\theta = \sec^2\theta - 1$:

$$I_{2n+1} = \int_0^{\frac{1}{4}\pi} \tan^{2n-1}\theta \, (-1 + \sec^2\theta) \, d\theta = -I_{2n-1} + \int_0^1 u^{2n-1} \, du = -I_{2n-1} + \frac{1}{2n}.$$

To evaluate the second integral, set $u = \tan\theta$ so that $du = \sec^2\theta \, d\theta$.

Repeating the process gives

$$I_{2n+1} = -I_{2n-1} + \frac{1}{2n} = I_{2n-3} - \frac{1}{2(n-1)} + \frac{1}{2n} = \cdots = (-1)^n I_1 + \frac{1}{2n} - \frac{1}{2(n-1)} + \cdots + (-1)^{n+1} \frac{1}{2}.$$

The above sum (starting with $1/(2n)$) is the same as that in (†) overleaf, so it only remains to evaluate I_1, corresponding to $n = 0$ in $(**)$:

$$I_1 = \int_0^{\frac{1}{4}\pi} \tan\theta \, d\theta = -\ln(\cos\theta)\Big|_0^{\frac{1}{4}\pi} = -\ln(1/\sqrt{2}) = \tfrac{1}{2}\ln 2,$$

as required.

We deduce the expression (‡) overleaf for $\ln 2$ using the first line of the question. When n is very large, I_{2n+1} is very small, being the area under a graph which is almost zero for almost all of the range of integration. In the limit $n \to \infty$, we set $I_{2n+1} = 0$ in (†) which leads immediately to (‡).

(ii) To obtain the formula for $\frac{1}{4}\pi$, we follow the above method using I_{2n} instead of I_{2n+1}. This time we have to calculate I_0:

$$I_0 = \int_0^{\frac{1}{4}\pi} 1 \, d\theta = \frac{\pi}{4}.$$

Post-mortem

You might think that the method of obtaining the formula for $\frac{1}{4}\pi$ in this question is rather indirect; one could instead just integrate the formula

$$\frac{d\tan^{-1} x}{dx} = \frac{1}{1+x^2} = 1 - x^2 + x^4 - \cdots \qquad (*)$$

term by term and get the result immediately by setting $x = 1$. The virtue of the method used in the question is that it gives an explicit form (an integral) of the remainder after n terms of the series. We were able to show, by means of a sketch, that the remainder tends to zero as n tends to infinity; in other words, we showed that the series converges. Although the sketch method of proof is a bit crude, it can easily be made more rigorous once the concept of integration is more carefully defined. On the other hand, integrating $(*)$ and setting $x = 1$ is a bit delicate, since the series only converges for $x^2 < 1$.

Problem 23: Intersection of ellipses (✓)

> Show that the equation of any circle passing through the points of intersection of the ellipse
>
> $$(x+2)^2 + 2y^2 = 18$$
>
> and the ellipse
>
> $$9(x-1)^2 + 16y^2 = 25$$
>
> can be written in the form
>
> $$x^2 - 2ax + y^2 = 5 - 4a \,.$$

2002 Paper I

Comments

You don't need to know anything about ellipses to do this question.

When this question was set it seemed too easy. But quite a high proportion of the candidates made no real progress. Of course, it is not easy to keep a cool head under examination conditions; but surely it is obvious that either x or y has to be eliminated from the two equations for the ellipses; and having decided that, it is obvious which one to eliminate.

It is worth thinking about how many points of intersection of the ellipses we are expecting: a sketch might help if you know what shapes the ellipses are.

Solution to problem 23

First we have to find the intersections of the two ellipses by solving the simultaneous equations

$$(x+2)^2 + 2y^2 = 18 \tag{1}$$
$$9(x-1)^2 + 16y^2 = 25 . \tag{2}$$

We can eliminate y by multiplying equation (1) by 8 and subtracting equation (2), so that at the intersections

$$8(x+2)^2 - 9(x-1)^2 = 144 - 25$$

i.e.

$$x^2 - 50x + 96 = 0$$

i.e.

$$(x-2)(x-48) = 0 .$$

The two possible values for x at the intersections are therefore 2 and 48.

Next we find the values of y at the intersection. Taking $x = 2$ and substituting into equation (1) gives $16 + 2y^2 = 18$, so $y = \pm 1$. Taking $x = 48$ gives $50^2 + 2y^2 = 18$ which has no (real) roots. Thus there are two points of intersection, at $(2, 1)$ and $(2, -1)$.

Now we go after the circle. Suppose that a circle through the points of intersection has centre (p, q) and radius R. Then the equation of the circle is

$$(x-p)^2 + (y-q)^2 = R^2 .$$

Setting $(x, y) = (2, 1)$ and $(x, y) = (2, -1)$ gives two equations:.

$$(2-p)^2 + (1-q)^2 = R^2 , \quad (2-p)^2 + (-1-q)^2 = R^2 .$$

Subtracting the two equations gives $q = 0$ (it is obvious anyway, because of the symmetry of both ellipses under reflections in the x axis, that the centre of the circle must lie on the y axis). Thus the equation of any circle passing through the intersections is

$$(x-p)^2 + y^2 = (2-p)^2 + 1 ,$$

which simplifies to the given result with $p = a$.

Post-mortem

As commented earlier, there really wasn't much to this question apart from solving simultaneous equations and quadratic equations. I suppose that the daunting feature is that very little help is given in the way of intermediate steps of answers.

Another daunting feature is the unexplained parameter a in the equation of the circle. Two ellipses can intersect in four points, three points (if the ellipses touch rather than intersect at one of the points), two points, one point or not at all.

In general, we wouldn't expect to be able to draw a circle through four given points. We would expect exactly one circle through three given points (not lying on a line), but a whole family of circles through any two points. Our ellipses intersect in two points, which is why the circle depends on a parameter a.

Problem 24: Sketching $x^m(1-x)^n$ (✓✓✓)

Let $f(x) = x^m(x-1)^n$, where m and n are both integers greater than 1. Show that

$$f'(x) = \left(\frac{m}{x} - \frac{n}{1-x}\right) f(x).$$

Show that the curve $y = f(x)$ has a stationary point in the interval $0 < x < 1$. By considering $f''(x)$, show that this stationary point is a maximum if n is even and a minimum if n is odd. Sketch the graphs of $f(x)$ in the four cases that arise according to the values of m and n.

2002 Paper I

Comments

There is quite a lot in this question, but it is one of the best STEP questions on basic material that I came across.

First, you have to find $f'(x)$. The very first part (giving a form of $f'(x)$) did not occur in the original STEP question. This particular expression is extremely helpful when if comes to finding the value of $f''(x)$ at the stationary point. In the actual exam, a handful of candidates successfully found a general expression for $f''(x)$ in terms of x then evaluated it at the stationary point: first class work.

You should find that the value of $f''(x)$ at the stationary point can be expressed in the form $(\cdots)f(x)$, where the factor in the brackets is quite simple and always negative, so that the nature of the stationary point depends only on the sign of $f(x)$. Actually, this is intuitively obvious: $f(0) = f(1) = 0$ so the one stationary point with $0 < x < 1$ must be, for example, a maximum if $f(x) > 0$ for $0 < x < 1$.

For the sketches, you really have only to think about the behaviour of $f(x)$ when $|x|$ is large and the sign of $f(x)$ between $x = 0$ and $x = 1$. That allows you to piece together the graph, knowing that there is only one stationary point between $x = 0$ and $x = 1$. You should then be able to identify the four cases referred to.

Note that, very close to $x = 0$, $f(x) \approx x^m(-1)^n$ so it is easy to see how the graph there depends on n and m; you can use this to check that you have got the graphs right.

Post-mortem

Don't read this until you have worked through the question!

You may have noticed a little carelessness in the first line of the solution: what happens in the logarithmic differentiation if any of $f(x)$ or x or $x-1$ are negative? The answer is that it doesn't matter. One way to deal with the problem of logs with negative arguments is to put modulus signs everywhere using the correct result

$$\frac{d}{dx} \ln |f(x)| = \frac{f'(x)}{f(x)}.$$

If you are not sure of this, try it on $f(x) = x$ taking the two cases $x > 0$ and $x < 0$ separately.

A more sophisticated way of dealing with logs with negative arguments is to note that $\ln(-f(x)) = \ln f(x) + \ln(-1)$. We don't have to worry about $\ln(-1)$ because it is a constant (of some sort) and so won't affect the differentiation. Actually, a value of $\ln(-1)$ can be obtained by taking logs of Euler's famous formula $e^{i\pi} = -1$ giving $\ln(-1) = i\pi$.

Solution to problem 24

The first result can be established by differentiating $f(x)$ directly, but the neat way to do it is to start with $\ln f(x)$:

$$\ln f(x) = m\ln x + n\ln(x-1) \implies \frac{f'(x)}{f(x)} = \frac{m}{x} + \frac{n}{x-1} \implies f'(x) = \left(\frac{m}{x} - \frac{n}{1-x}\right) f(x),$$

as required. Writing this as $f'(x) = mx^{m-1} - n(x-1)^{n-1}$, we see that $f(x)$ has stationary points at $x = 0$ and $x = 1$ (since $m-1 > 0$ and $n-1 > 0$) and when

$$\frac{m}{x} - \frac{n}{1-x} = 0,$$

i.e. when $m(1-x) - nx = 0$. Solving this last equation for x gives $x = \dfrac{m}{m+n}$, which lies between 0 and 1 since m and n are positive.

Next we calculate $f''(x)$. Starting with

$$f'(x) = \left(\frac{m}{x} - \frac{n}{1-x}\right) f(x),$$

we obtain

$$f''(x) = \left(-\frac{m}{x^2} - \frac{n}{(1-x)^2}\right) f(x) + \left(\frac{m}{x} - \frac{n}{1-x}\right) f'(x).$$

At a stationary point, the second of these two terms is zero because $f'(x) = 0$, leaving

$$f''(x) = \left(-\frac{m}{x^2} - \frac{n}{(x-1)^2}\right) f(x).$$

The bracketed expression is negative so at a stationary point $f''(x) < 0$ if $f(x) > 0$ and $f''(x) > 0$ if $f(x) < 0$.

The sign of $f(x)$ for $0 < x < 1$ is the same as the sign of $(x-1)^n$, since $x^m > 0$ when $x > 0$. For the stationary point in the interval $0 < x < 1$, $(x-1)^n > 0$ if n is even and $(x-1)^n < 0$ if n is odd. Thus $f''(x) < 0$ if n is even and $f''(x) > 0$ if n is odd, which is the required result.

The four cases to sketch are determined by whether m and n are even or odd. The easiest way to understand what is going on is to consider the graphs of x^m and $(x-1)^n$ separately, then try to join them up at the stationary point between 0 and 1. If m is odd, then $x = 0$ is a point of inflection, but if m is even it is a maximum or minimum according to the sign of $(x-1)^n$. You should also think about the behaviour for large $|x|$. All the various bits of information (including the nature of the turning point investigated above) should all piece neatly together.

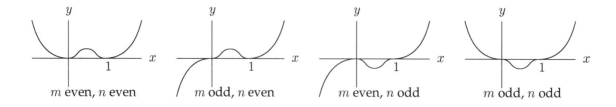

m even, n even $\qquad$ m odd, n even $\qquad$ m even, n odd $\qquad$ m odd, n odd

Problem 25: Inequalities by area estimates (✓✓✓)

> Give a sketch of the curve $y = \dfrac{1}{1+x^2}$, for $x \geqslant 0$.
>
> Find the equation of the line that intersects the curve at $x = 0$ and is tangent to the curve at some point with $x > 0$. Prove that there are no further intersections between the line and the curve. Draw the line on your sketch.
>
> By considering the area under the curve for $0 \leqslant x \leqslant 1$, show that $\pi > 3$.
>
> By considering the volume formed by rotating the curve about the y axis, show also that $\ln 2 > \frac{2}{3}$.
>
> **Note:** $\displaystyle\int_0^1 \dfrac{1}{1+x^2}\,\mathrm{d}x = \dfrac{\pi}{4}$.

2002 Paper I

Comments

There is quite a lot to this question, so ✓✓✓ even though none of it is particularly difficult (very surprising to find it on Paper I). The most common mistake in finding the equation of the tangent is to muddle the y and x that occur in the equation of the line ($y = mx + c$) with the coordinates of the point at which the tangent meets the curve, getting 'constants' m and c that depend on x. I'm sure you wouldn't make this rather elementary mistake normally, but it is surprising what people do under examination conditions.

The reason for sketching the curve lies in the last parts: the shape of the curve relative to the straight line provides the inequality.

The note at the end of the question had to be given because integrals of that form (giving inverse trigonometric functions) are not in the core A-level syllabus.

Solution to problem 25

Your sketch should show a graph which has gradient 0 at $(0,1)$ and which asymptotes to the x-axis for large x. The extra line in my sketch is the tangent to the graph from the point $(0,1)$ which is required later.

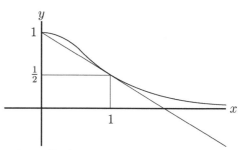

First we find the equation of the tangent to the curve $y = (1+x^2)^{-1}$ at the point (p,q). The gradient of the curve at the point (p,q) is $-2p(1+p^2)^{-2}$, i.e. $-2pq^2$, so the equation of the tangent is

$$y = -2pq^2 x + c$$

where c is given by $q = -2pq^2 p + c$. This line is supposed to pass through the point $(0,1)$, so $1 = c$. Thus $1 = q + 2p^2 q^2$, or (replacing q with $1/(1+p^2)$)

$$1 = \frac{1}{1+p^2} + \frac{2p^2}{(1+p^2)^2}$$

which simplifies to

$$(1+p^2)^2 = 1 + 3p^2 \quad \text{i.e.} \quad p^4 = p^2 \,.$$

The only positive solution is $p = 1$, so the equation of the line is $y = -\frac{1}{2}x + 1$.

We can check that this line does not meet the curve again by solving the equation

$$-\frac{x}{2} + 1 = \frac{1}{1+x^2} \,.$$

Multiplying by $(1+x)^2$ gives

$$(1 - \tfrac{1}{2}x)(1+x^2) = 1 \quad \text{i.e.} \quad x^3 - 2x^2 + x = 0 \,.$$

Factorising shows that $x = 1$ and $x = 0$ satisfy this equation (these are the known roots, $x = 1$ being a double root corresponding to the tangency) and that there are no more roots.

The area under the curve for $0 \leqslant x \leqslant 1$ is $\frac{1}{4}\pi$ as given. The sketch shows that this area is greater than the area under the tangent line for $0 \leqslant x \leqslant 1$, which is $\frac{1}{2} + \frac{1}{4}$ (the area of the rectangle plus the area of the triangle above it). Comparing with with $\frac{1}{4}\pi$ gives the required result.

The volume formed by rotating the curve about the y axis is

$$\int_0^1 2\pi y x \, dx = \pi \int_0^1 \frac{2x}{1+x^2} \, dx = \pi \ln 2 \,.$$

This is greater than the volume formed by rotating the line about the y axis, which is

$$\int_0^1 2\pi y x \, dx = 2\pi \int_0^1 (1 - \tfrac{1}{2}x) x \, dx = \tfrac{2}{3}\pi \,.$$

Comparison gives the required result.

Post-mortem

As I said before, no step of this question is particularly difficult, but *lots* of different ideas are required. The last part is very challenging, because you have to do and compare two separate volume integrals without any intermediate steer at all. You should be very pleased if you made good progress with it.

The important lesson to learn from this is that perseverance will eventually pay off.

Problem 26: Simultaneous integral equations (✓✓)

Let
$$I = \int_0^a \frac{\cos x}{\sin x + \cos x}\, dx \quad \text{and} \quad J = \int_0^a \frac{\sin x}{\sin x + \cos x}\, dx,$$
where $0 \leqslant a < \tfrac{3}{4}\pi$. By considering $I + J$ and $I - J$, show that $2I = a + \ln(\sin a + \cos a)$.

Find also:

(i) $\displaystyle\int_0^{\frac{1}{2}\pi} \frac{\cos x}{p \sin x + q \cos x}\, dx$, where p and q are positive numbers;

(ii) $\displaystyle\int_0^{\frac{1}{2}\pi} \frac{\cos x + 4}{3 \sin x + 4 \cos x + 25}\, dx.$

2002 Paper I

Comments

This is a model for a perfect STEP question: you are told how to do the first part, and you have to adapt the idea on your own for the later parts. Note the structure of the question: there is a 'stem' (the first paragraph) containing material that will be useful for both the later parts.

In the examination, candidates who were successful in parts (i) and (ii) nearly always started off with the statement 'Now let $I = \cdots$ and $J = \cdots$'. If you are stuck with part (ii), the choice of significant numbers (3, 4 and 25) should provide a clue.

When you arrive at an answer for part (i), you will of course check that it agrees with the given result in the opening paragraph when $a = \tfrac{1}{2}\pi$ and $p = q = 1$.

You will no doubt have noticed the restriction $0 \leqslant a < \tfrac{3}{4}\pi$ given in the first paragraph (and also the restrictions on p and q in part (i)). You should try to work out its purpose because it might provide an insight into the method of tackling the question, though in this case it doesn't.

Solution to problem 26

For the first part, we regard the two integrals essentially as a pair of simultaneous equations, adding and subtracting to simplify them. We have

$$I + J = \int_0^a \frac{\cos x + \sin x}{\sin x + \cos x} dx = \int_0^a dx = a$$

$$I - J = \int_0^a \frac{\cos x - \sin x}{\sin x + \cos x} dx = \ln(\cos a + \sin a)$$

(in the second integral, the numerator is the derivative of the denominator). Adding these equations gives the required expression for $2I$.

(i) Similarly, let $I = \int_0^{\frac{1}{2}\pi} \frac{\cos x}{p \sin x + q \cos x} dx$ and $J = \int_0^{\frac{1}{2}\pi} \frac{\sin x}{p \sin x + q \cos x} dx$. Then

$$qI + pJ = \int_0^{\frac{1}{2}\pi} \frac{q \cos x + p \sin x}{p \sin x + q \cos x} dx = \tfrac{1}{2}\pi$$

$$pI - qJ = \int_0^{\frac{\pi}{2}} \frac{p \cos x - q \sin x}{p \sin x + q \cos x} dx = \ln(p \sin \tfrac{1}{2}\pi + q \cos \tfrac{1}{2}\pi) - \ln(p \sin 0 + q \cos 0) = \ln \frac{p}{q}.$$

Now we solve these two equations simultaneously for I:

$$(p^2 + q^2)I = \frac{q\pi}{2} + p \ln \frac{p}{q}.$$

(ii) This time, let $I = \int_0^{\frac{1}{2}\pi} \frac{\cos x + 4}{3 \sin x + 4 \cos x + 25} dx$ and $J = \int_0^{\frac{1}{2}\pi} \frac{\sin x + 3}{3 \sin x + 4 \cos x + 25} dx$. Then

$$4I + 3J = \int_0^{\frac{1}{2}\pi} \frac{4 \cos x + 3 \sin x + 25}{3 \sin x + 4 \cos x + 25} dx = \frac{\pi}{2}$$

$$3I - 4J = \int_0^{\frac{1}{2}\pi} \frac{3 \cos x - 4 \sin x}{3 \sin x + 4 \cos x + 25} dx = \ln \tfrac{28}{29}.$$

Solving simultaneously gives

$$25I = 2\pi + 3 \ln \tfrac{28}{29}.$$

Post-mortem

I very much like this question. The first part appeared on STEP Paper I in 1995 (the second part of the 1995 question was an integral completely unrelated to the first part — that wouldn't happen now) and I was completely taken by surprise.

In order to use it again for STEP in 2002, I added parts (i) and (ii). It took me some time to think of a suitable extension. I was disappointed to find that the basic idea is more or less a one-off: there are very few denominators, besides the ones given, that lead to integrands amenable to this trick. But I was pleased with what I came up with. The question leads you through the opening paragraph and the extra parts depend very much on your having understood why the opening paragraph works. The reason for restriction $0 \leqslant a < 3\pi/4$ is that the denominator should not be 0 for any value of x in the range of integration — otherwise, the integral is undefined. Writing $\sin x + \cos x = \sqrt{2} \sin(x + \pi/4)$ shows that the denominator is first zero when $x = 3\pi/4$.

You might like to think how you would evaluate these integrals without using the trick method of this question. Perhaps the easiest way is to use the substitution $t = \tan(\tfrac{1}{2}x)$ which converts the denominator to a quadratic in x.

Problem 27: Relation between coefficients of quartic for real roots (✓✓)

In this question you may assume that, if $k_1, \ldots, k_n$ are distinct positive real numbers, then

$$\frac{1}{n}\sum_{r=1}^{n} k_r > \left(\prod_{r=1}^{n} k_r\right)^{\frac{1}{n}},$$

i.e. their arithmetic mean is greater than their geometric mean.

Suppose that a, b, c and d are positive real numbers such that the polynomial

$$f(x) = x^4 - 4ax^3 + 6b^2x^2 - 4c^3x + d^4$$

has four distinct positive roots.

(i) By considering the relationship between the coefficients of f and its roots, show that $c > d$.

(ii) By differentiating f, show that $b > c$.

(iii) Show that $a > b$.

1997 Paper III

Comments

This result is both surprising and pleasing.

The question looks difficult, but you don't have to go very far before you come across something to substitute into the given arithmetic mean/geometric mean (AM/GM) inequality.

To obtain the relationship between the coefficients and the roots for part (i), you need to write the quartic equation in the form $(x-p)(x-q)(x-r)(x-s) = 0$.

Watch out for the condition in the given form of the arithmetic/geometric inequality that the numbers are distinct: you will have to show that any numbers (or algebraic expressions) you use in the inequality are distinct.

I puzzled over this question for ages, not understanding the idea behind the question[20]

It will be clear from the proof that it can be generalised to any equation of the form

$$x^N + \sum_{k=0}^{N-1} \binom{N}{k}(-a_k)^{N-k} x^k = 0,$$

where the numbers a_k are distinct and positive, which has positive distinct roots.

Eventually, I realised that the result of the question has really nothing to do with quartic equations, though quartic equations provide a neat method of proof. The inequalities, which apply to any positive numbers, form a sequence interpolating between the AM and the GM. These inequalities were first derived by Newton and Maclaurin in the 17th century but don't seem to be very well known now.

---

[20] I mentioned this in a talk I gave to sixth formers, and later saw irritatingly on www.thestudentroom.co.uk a comment to the effect that 'even Dr Siklos can't do this STEP question'. Of course I could *do* it; I wanted to *understand* it, as I hope you do too.

Solution to problem 27

(i) First write
$$f(x) \equiv (x-p)(x-q)(x-r)(x-s)$$
where p, q, r and s are the four roots of the equation (known to be real, positive and distinct). Multiplying out the brackets and comparing with $x^4 - 4ax^3 + 6b^2x^2 - 4c^3x + d^4$ shows that $pqrs = d^4$ and $pqr + qrs + rsp + spq = 4c^3$.

The required result, $c > d$, follows immediately by applying the AM/GM inequality to the positive real numbers pqr, qrs, rsp and spq:
$$c^3 = \frac{pqr + qrs + rsp + spq}{4} > [(pqr)(qrs)(rsp)(spq)]^{1/4} = [p^3q^3r^3s^3]^{1/4} = [d^{12}]^{1/4}.$$

Taking the cube root (c and d are positive) preserves the inequality.

The AM/GM inequality at the beginning of the question is stated only for the case when the numbers are distinct (though in fact it holds provided at least two of the numbers are distinct). To use the inequality as above, we must therefore show that no two of pqr, qrs, rsp and spq are equal. This follows immediately from the fact that the roots are distinct and non-zero. For if, for example, $pqr = qrs$ then $qr(p-s) = 0$ which means that $q = 0$, $r = 0$ or $p = s$ all of which are ruled out.

(ii) The polynomial $f'(x)$ is cubic so it has three zeros (roots). These are at the turning points of $f(x)$, which lie between the zeros of $f(x)$ and are therefore distinct and positive.

Now
$$f'(x) = 4x^3 - 12ax^2 + 12b^2x - 4c^3,$$
so at the turning points
$$x^3 - 3ax^2 + 3b^2x - c^3 = 0.$$

Suppose that the roots of this cubic equation are u, v and w all real, distinct and positive. Then comparing $x^3 - 3ax^2 + 3b^2x - c^3 = 0$ and $(x-u)(x-v)(x-w)$ shows that $c^3 = uvw$ and $3b^2 = vw + wu + uv$.

The required result, $b > c$, follows immediately by applying the AM/GM inequality to the positive distinct numbers vw, wu and uv.

(iii) Apply similar arguments to $f''(x)/12$.

Post-mortem

The general inequalities mentioned in the comments above are interesting. For example, for four numbers a_1, a_2, a_3 and a_4, we define
$$S_1 = \frac{a_1 + a_2 + a_3 + a_4}{4}, \quad S_2 = \frac{a_2a_3 + a_3a_1 + a_1a_2 + a_4a_1 + a_4a_2 + a_4a_3}{6},$$
$$S_3 = \frac{a_2a_3a_4 + a_1a_3a_4 + a_1a_2a_4 + a_1a_2a_3}{4} \quad \text{and} \quad S_4 = a_1a_2a_3a_4.$$

Except for the denominators and alternating signs, S_i is the coefficient of x^{4-i} in the equation $(x-a_1)(x-a_2)(x-a_3)(x-a_4) = 0$, but that doesn't matter. Maclaurin's inequalities are
$$S_1 \geqslant S_2^{\frac{1}{2}} \geqslant S_3^{\frac{1}{3}} \geqslant S_4^{\frac{1}{4}}.$$

What is really surprising is that we proved the three stronger inequalities from the apparently weaker AM>GM inequality ($S_1 \geqslant S_4^{\frac{1}{4}}$): a rare example of pulling yourself up by your bootstraps.

Problem 28: Fermat numbers $(\checkmark\checkmark)$

The nth Fermat number, F_n, is defined by

$$F_n = 2^{2^n} + 1, \qquad n = 0, 1, 2, \ldots,$$

where 2^{2^n} means 2 raised to the power 2^n. Calculate F_0, F_1, F_2 and F_3. Show that, for $k = 1$, $k = 2$ and $k = 3$,

$$F_0 F_1 \ldots F_{k-1} = F_k - 2. \qquad (*)$$

Prove, by induction, or otherwise, that $(*)$ holds for all $k \geqslant 1$. Deduce that no two Fermat numbers have a common factor greater than 1.
Hence show that there are infinitely many prime numbers.

2002 Paper II

Comments

Fermat (1601–1665) conjectured that every number of the form $2^{2^n} + 1$ is a prime number. F_0 to F_4 are indeed prime, but Euler showed in 1732 that F_5 (4294967297) is divisible by 641. As can be seen, Fermat numbers get very big and not many more have been investigated; but those that have been investigated have been found not to be prime numbers. It is now conjectured that in fact only a finite number of Fermat numbers are prime numbers.

The Fermat numbers have a geometrical significance as well: Gauss proved that a regular polygon of n sides can be inscribed in a circle using a Euclidean construction (i.e. only a straight edge and a compass) if and only if n is a power of 2 times a product of distinct Fermat primes.

The very last part of this question, showing that the number of primes is infinite, is completely unexpected and delightful. It comes from *Proofs from THE BOOK*[21], a set of proofs thought by Paul Erdös to be heaven-sent. Most of them are much less elementary this one. Erdös was an extraordinarily prolific mathematician. He had almost no personal belongings and no home[22]. He spent his life visiting other mathematicians and proving theorems with them. He collaborated with so many people that every mathematician is (jokingly) assigned an *Erdös number*; for example, if you wrote a paper with someone who wrote a paper with Erdös, you are given the Erdös number 2. Most mathematicians seem to have an Erdös number less than 8.

---

[21] M. Aigler and G.M. Ziegler (Springer, 1999). The idea behind the proof was given by Christian Goldbach (1690–1764), who is best known for his conjecture that there are an infinite number of *twin primes*, that is, prime numbers that are two apart such as 17 and 19.

[22] See *The Man Who Loved Only Numbers* by Paul Hoffman published by Little Brown and Company in 1999 — a wonderfully readable and interesting book.

Solution to problem 28

To begin with, we have by direct calculation that $F_0 = 3$, $F_1 = 5$, $F_2 = 17$ and $F_3 = 257$, so it is easy to verify (∗) for $k = 1, 2$ and 3.

For the induction, we start by assuming that the result holds for $k = m$ so that
$$F_0 F_1 \ldots F_{m-1} = F_m - 2 .$$

We need to show that this implies that the result holds for $k = m + 1$, i.e. that
$$F_0 F_1 \ldots F_m = F_{m+1} - 2 . \tag{∗}$$

Starting with the left hand side of this equation, we have
$$F_0 F_1 \ldots F_{m-1} F_m = (F_m - 2) F_m = (2^{2^m} - 1)(2^{2^m} + 1) = \left(2^{2^m}\right)^2 - 1 = 2^{2^{m+1}} - 1 = F_{m+1} - 2,$$

as required. Since we know that the result holds for $k = 1$, the induction is complete.

To show that no two Fermat number have a common factor (other than 1), we proceed by contradiction. Suppose p divides F_l and F_m, where $l < m$. Then p divides $F_0 F_1 \ldots F_l \ldots F_{m-1}$ and therefore p divides $F_m - 2$, by (∗), as well as F_m. But this is impossible: all Fermat numbers are odd so no number other than 1 divides both $F_m - 2$ and F_m. Hence no two Fermat numbers have a common factor greater than 1.

For the last part, note that every number can be written uniquely as a product of prime factors and that because the Fermat numbers are co-prime, each prime can appear in at most one Fermat number. Thus, since there are infinitely many Fermat numbers, there must be infinitely many primes.

Post-mortem

This question is largely about proof. The above solution has a proof by induction and a proof by contradiction. The last part could have been written as a proof by contradiction too.

The difficulty is in presenting the proof. It is not enough to understand your own proof: you have to be able to set it out so that it is clear to the reader. This is a vital part of mathematics and its most useful transferable skill.

At the meeting in which this question was first considered, one of the examiners said that he didn't see the point of the question, since there is already a well-known proof (due to Euclid, possibly) that there are infinitely many primes. As mentioned overleaf, this proof is due to Goldbach, and I could only respond that if Goldbach thought it worthwhile, that was good enough for me.

Problem 29: Telescoping series (✓✓)

(i) Show that the sum S_N of the first N terms of the series

$$\frac{1}{1.2.3} + \frac{3}{2.3.4} + \frac{5}{3.4.5} + \cdots + \frac{2n-1}{n(n+1)(n+2)} + \cdots$$

is

$$\frac{1}{2}\left(\frac{3}{2} + \frac{1}{N+1} - \frac{5}{N+2}\right).$$

What is the limit of S_N as $N \to \infty$?

(ii) The numbers a_n are such that

$$\frac{a_n}{a_{n-1}} = \frac{(n-1)(2n-1)}{(n+2)(2n-3)}.$$

Find an expression for $\dfrac{a_n}{a_1}$ and hence, or otherwise, evaluate $\sum_{n=1}^{\infty} a_n$ when $a_1 = \frac{2}{9}$.

1998 Paper II

Comments

If you haven't the faintest idea how to do the sum, then look at the first line of the solution; but don't look without first having had a long think about it, picking up ideas from the form of the given answer.

Limits form an important part of first year university mathematics. The definition of a limit is one of the basic ideas in *analysis*, which is the rigorous study of calculus. At the end of part (i), no such definition is needed: you just see what happens when N gets very large (some terms get very small and eventually go away).

Part (ii) looks as if it might be some new idea. Since this is STEP, you will probably realise that the new series must be closely related to the series in part (i). The peculiar choice for a_1 ($= \frac{2}{9}$) should make you suspect that the sum will come out to some nice round number (not in fact round in this case, but straight and thin).

Having decided how to do the first part, please don't use the 'cover up' rule unless you understand why it works: mathematics at this level is not a matter of applying learned recipes. See the post-mortem for more thoughts on this matter.

Solution to problem 29

The given answer to the sum suggests partial fractions. It is difficult to think of any other way of starting, so let's convert the general term of the series to partial fractions in the hope that something good might happen. Set

$$\frac{2n-1}{n(n+1)(n+2)} \equiv \frac{A}{n} + \frac{B}{n+1} + \frac{C}{n+2}, \qquad (*)$$

then use your favourite method to find $A = -\frac{1}{2}$, $B = 3$ and $C = -\frac{5}{2}$. Note that $A + B + C+$). The series can now be written

$$\left(\frac{A}{1} + \frac{B}{2} + \frac{C}{3}\right) + \left(\frac{A}{2} + \frac{B}{3} + \frac{C}{4}\right) + \left(\frac{A}{3} + \frac{B}{4} + \frac{C}{5}\right) +$$

$$+ \cdots + \left(\frac{A}{N-2} + \frac{B}{N-1} + \frac{C}{N}\right) + \left(\frac{A}{N-1} + \frac{B}{N} + \frac{C}{N+1}\right) + \left(\frac{A}{N} + \frac{B}{N+1} + \frac{C}{N+2}\right). \qquad (**)$$

Now we collect up terms with the same denominators and find that all the terms in the series cancel, except those with denominators $1, 2, N+1$ and $N+2$. These exceptions sum to the required answer. The limit as $N \to \infty$ is $\frac{3}{4}$ since the other two terms obviously tend to zero.

For part (ii), we note that $\dfrac{a_n}{a_{n-1}} = \dfrac{b_n}{b_{n-1}}$, where b_n is the general term of the series in part (i). Thus

$$\frac{a_n}{a_1} = \frac{b_n}{b_1} \quad \text{and} \quad \sum_1^\infty a_n = \frac{a_1}{b_1} \sum_1^\infty b_n = \frac{2/9}{1/6} \times \frac{3}{4} = 1.$$

Alternatively, we can write out the nth term explicitly:

$$a_n = \frac{(n-1)(2n-1)}{(n+2)(2n-3)} a_{n-1} = \frac{(n-1)(2n-1)}{(n+2)(2n-3)} \frac{(n-2)(2n-3)}{(n+1)(2n-5)} a_{n-2}$$

$$= \frac{(n-1)(2n-1)}{(n+2)(2n-3)} \frac{(n-2)(2n-3)}{(n+1)(2n-5)} \frac{(n-3)(2n-5)}{(n)(2n-7)} \cdots \frac{5 \times 11}{8 \times 9} \frac{4 \times 9}{7 \times 7} \frac{3 \times 7}{6 \times 5} \frac{2 \times 5}{5 \times 3} \frac{1 \times 3}{4 \times 1} a_1$$

$$= \frac{2n-1}{n(n+1)(n+2)} \frac{3 \times 2 \times 1}{1} a_1 = \frac{12}{9} \frac{2n-1}{n(n+1)(n+2)},$$

all other terms cancelling. Now using the result of the first part gives $\sum_1^\infty a_n = 1$.

Post-mortem

A small point of technique: equation $(**)$ was made much clearer (and it saved writing) to stick with A, B and C instead of using $-\frac{1}{2}$, 3 and $-\frac{5}{2}$. The method would not depend on the arithmetic values of these constants.

There are various methods for finding A, B and C in $(*)$.

One is to set $n = 1$, $n = 2$ and $n = 3$ consecutively and obtain three simultaneous equations.

Another is to multiply up and simplify, giving $2n - 1 = (A + B + C)n^2 + (3A + 2B + C)n + 2A$. You then equate coefficients of powers of n.

A better way is to multiply up without simplifying, giving $2n - 1 = A(n+1)(n+2) + Bn(n+2) + Cn(n+1)$. You then choose values for n that give quick results: for example, setting $n = -1$ gives $B = 3$ immediately. This is of course the method behind the iniquitous 'cover-up rule'. Note that the 'equivalence' sign, $\equiv$, indicates an identity (something that holds for all values of n) rather than an equation to solve for n, so n doesn't have to be a positive integer (you could set $n = -\frac{1}{2}$ if you fancied it).

Problem 30: Integer solutions of cubics (✓✓✓)

(i) Show that, if m is an integer such that
$$(m-3)^3 + m^3 = (m+3)^3, \qquad (*)$$
then m^2 is a factor of 54. Deduce that there is no integer m which satisfies the equation $(*)$.

(ii) Show that, if n is an integer such that
$$(n-6)^3 + n^3 = (n+6)^3, \qquad (**)$$
then n is even. By writing $n = 2m$ deduce that there is no integer n which satisfies the equation $(**)$.

(iii) Show that, if n is an integer such that
$$(n-a)^3 + n^3 = (n+a)^3, \qquad (***)$$
where a is a non-zero integer, then n is even and a is even. Deduce that there is no integer n which satisfies the equation $(***)$.

1998 Paper II

Comments

I slightly simplified the first two parts of the question, which comprised the whole of the original STEP question, and added the third part. This last part is conceptually tricky and very interesting: hence the ✓✓✓ rating.

It is a fairly standard technique in this sort of number theoretic problem to investigate whether the equation *balances*: for example, if the left hand side is even, then the right hand side must also be even. If this fails, then the equation can have no solutions.

Solution to problem 30

(i) Simplifying gives $m^2(m-18) = 54$.

Both m^2 and $(m-18)$ must divide 54, which is impossible since the only squares that divide 54 are 1 and 9, and neither $m = 1$ nor $m = 3$ satisfies $m^2(m-18) = 54$.

You could also argue that $m - 18$ must be positive so $m \geqslant 19$ and $m^2 \geqslant 361$ which is a contradiction.

(ii) The easiest method is a proof by contradiction. Suppose therefore that n is odd. The two terms on the left hand side are both odd, which means that the left hand side is even. But the right hand side is odd so the equation cannot balance if n is odd.

Setting $n = 2m$ gives
$$(2m-6)^3 + (2m)^3 = (2m+6)^3.$$

Taking a factor of 2^3 out of each term leaves $(m-3)^3 + m^3 = (m+3)^3$ which is the same as the equation that was shown in part (i) to have no solutions.

(iii) First we show that n is even, dealing with the cases a odd and a even separately. The first case is n odd and a odd. In that case, $(n-a)^3$ is even, n^3 is odd and $(n+a)^3$ is even, so the equation does not balance. In the second case, n is odd and a is even, the three terms are all odd and again the equation does not balance. Therefore n cannot be odd.

Next, we investigate the case n is even and a odd. This time the three terms are odd, even and odd respectively so there is no contradiction. But multiplying out the brackets and simplifying gives
$$n^2(n-6a) = 2a^3 \,.$$

Since n is even the left hand side is divisible by 4, because of the factor n^2. That means that a^3 is divisible by 2 which cannot be the case since a is odd.

Now that we know n and a are both even, we can follow the method used in part (ii) and set $n = 2m$ and $a = 2b$. This gives
$$(2m-2b)^3 + (2m)^3 = (2m+2b)^3$$

from which a factor of 2^3 can be cancelled from each term. Thus m and b satisfy the same equation as n and a. They are therefore both even and we can repeat the process.

Repeating the process again and again will eventually result in an integer that is odd which will therefore not satisfy the equation that it is supposed to satisfy: a contradiction. There is therefore no integer n that satisfies $(***)$.

Post-mortem

The last proof is an example of what is known as the *method of infinite descent*. It was used by Fermat (1601 – 1665) to prove special cases of his Last Theorem[23], which of course is exactly what you are doing in this question. The method was probably invented by him and his faith in it sometimes led him astray. It is even possible that he thought he could use it to prove his Last Theorem in full. In fact, the proof of this theorem was only given in 1994 by Andrew Wiles; and it is 150 pages of pretty incomprehensible modern mathematics.

---

[23] The theorem says that the equation $x^n + y^n = z^n$, where x, y, z and n are positive integers, can only hold if $n = 1$ or 2

Problem 31: The harmonic series (✓✓✓)

The function f satisfies $0 \leqslant f(t) \leqslant K$ when $0 \leqslant t \leqslant 1$. Explain by means of a sketch, or otherwise, why

$$0 \leqslant \int_0^1 f(t)\, dt \leqslant K\,.$$

By considering $\int_0^1 \dfrac{t}{n(n-t)}\, dt$ or otherwise show that, if $n > 1$, then

$$0 \leqslant \ln\left(\frac{n}{n-1}\right) - \frac{1}{n} \leqslant \frac{1}{n-1} - \frac{1}{n}$$

and deduce that

$$0 \leqslant \ln N - \sum_{n=2}^{N} \frac{1}{n} \leqslant 1\,.$$

Deduce that $\displaystyle\sum_{n=1}^{N} \frac{1}{n} \to \infty$ as $N \to \infty$.

Noting that $2^{10} = 1024$, show also that if $N < 10^{30}$ then $\displaystyle\sum_{n=1}^{N} \frac{1}{n} < 101$.

1999 Paper I

Comments

Quite a lot of different ideas are required for this question; hence ✓✓✓.

The first hurdle is to decide what the constant K in the first part has to be in order to make the second part work. You might try to maximise the integrand using calculus, but that would be the wrong thing to do (you should find if you do this that the integrand has no turning point in the given range: it increases throughout the range).

The next hurdle is the third displayed equation, which follows from the preceding result. Then there is still more work to do.

It is not at all obvious that the series $\displaystyle\sum_{1}^{N} \frac{1}{n}$ (which is called the *harmonic series*) tends to infinity as N increases. There are easier ways of proving this than the method used in this question but this way tells us two interesting things. First it tells us that the sum increases very slowly indeed: the first 10^{30} terms only get to 100. Second it tells us that

$$0 \leqslant \sum_{1}^{N} \frac{1}{n} - \ln N \leqslant 1$$

for all N. (This is the third displayed equation in the question slightly rewritten.) Not only does the sum diverge, it does so logarithmically. In fact, in the limit $N \to \infty$,

$$\sum_{1}^{N} \frac{1}{n} - \ln N \to \gamma$$

where γ is *Euler's constant*. Its value is about $\tfrac{1}{2}$.

Solution to problem 31

Any sketch showing a squiggly curve all of which lies beneath the line $x = K$ and above the x-axis will do: area under curve (the integral) is less than the area of the rectangle (K).

Let's start the next part in the obvious way, by evaluating the integral.

Noting that $\dfrac{t}{n-t} = \dfrac{n}{(n-t)} - 1$, we have:

$$\int_0^1 \frac{t}{n(n-t)}\,dt = \int_0^1 \left(\frac{1}{n-t} - \frac{1}{n}\right) dt = \ln\left(\frac{n}{n-1}\right) - \frac{1}{n}.$$

The largest value of the integrand in the interval $[0,1]$ occurs at $t = 1$, since the numerator increases as t increases and the denominator decreases as t increases (remember that $n > 1$). Using the result of the first paragraph gives

$$0 \leqslant \ln\left(\frac{n}{n-1}\right) - \frac{1}{n} \leqslant \frac{1}{n(n-1)} = \frac{1}{n-1} - \frac{1}{n}.$$

Summing both sides from 2 to N and cancelling lots of terms in pairs gives

$$0 \leqslant \ln N - \sum_2^N \frac{1}{n} \leqslant 1 - \frac{1}{N}. \qquad (*)$$

Note that $1 - \dfrac{1}{N} < 1$. Since $\ln N \to \infty$ as $N \to \infty$, so also must $\displaystyle\sum_2^N \frac{1}{n}$ (it differs from the logarithm by less than 1).

Finally, rearranging the first inequality in $(*)$ gives $\displaystyle\sum_2^N \frac{1}{n} \leqslant \ln N$, i.e. $\displaystyle\sum_1^N \frac{1}{n} \leqslant \ln N + 1$. Setting $N = 10^{30}$, and using the inequalities $1000 < 1024$ and $e > 2$ (or $\ln 2 < 1$) gives:

$$\sum_1^{10^{30}} \frac{1}{n} \leqslant \ln(10^{30}) + 1 < \ln(1024^{10}) + 1 = 100 \ln 2 + 1 < 100 \ln e + 1 = 100 + 1.$$

Post-mortem

This question seems very daunting at first, because you are asked to prove a sequence of completely unfamiliar results. However, you should learn from this question that if you keep cool and follow the hints, explicit and implicit, you can achieve some surprisingly sophisticated results. You might think that this situation is very artificial: in real life, you do not receive hints to guide you. But often in mathematical research, hints *are* buried deep in the problem if only you can recognise them.

Problem 32: Integration by substitution (✓✓)

Find $\dfrac{dy}{dx}$ if
$$y = \frac{ax+b}{cx+d}.\qquad (*)$$

By using changes of variable of the form $(*)$, or otherwise, show that

$$\int_0^1 \frac{1}{(x+3)^2} \ln\left(\frac{x+1}{x+3}\right) dx = \frac{1}{6}\ln 3 - \frac{1}{4}\ln 2 - \frac{1}{12},$$

and evaluate the integrals

$$\int_0^1 \frac{1}{(x+3)^2} \ln\left(\frac{x^2+3x+2}{(x+3)^2}\right) dx \quad \text{and} \quad \int_0^1 \frac{1}{(x+3)^2} \ln\left(\frac{x+1}{x+2}\right) dx.$$

1999 Paper II

Comments

You will find that the change of variable in each case is clearly signalled: it is really only the denominator of $(*)$ that matters.

For the first integral, you do the obvious thing, but for the second and third integrals you have to be quite ingenious to get the argument of the logs in a suitable form. Once you have got the idea for the second integral, you should be able to see the connection with the third integral, but it would be hard to do the third integral without having done the second. That's what I like about this question: one thing leads to another.

I have written below the version of this question that was proposed by the setter, because I thought you would be interested to see how a question evolves.

By changing to the variable y defined by
$$y = \frac{2x-3}{x+1},$$
evaluate the integral
$$\int_2^4 \frac{2x-3}{(x+1)^3} \ln\left(\frac{2x-3}{x+1}\right) dx.$$

Evaluate the integral
$$\int_9^{25} \left(2z^{-\frac{3}{2}} - 5z^{-2}\right) \ln\left(2 - 5z^{-\frac{1}{2}}\right) dz.$$

Note in particular the way that the ideas in the final draft are closely knit and better structured: the first change of variable is strongly signalled but not explicit and the following parts, though based on the same idea, require increasing ingenuity. The first draft required quite a jump to evaluate the second integral.

Note also that the final integral of the first draft has an unpleasant contrived appearance, whereas the the integrals of the final draft are rather pleasing: beauty matters to mathematicians.

Solution to problem 32

Differentiating gives $\dfrac{dy}{dx} = \dfrac{a(cx+d) - c(ax+b)}{(cx+d)^2} = \dfrac{ad-bc}{(cx+d)^2}$.

For the first integral, set $y = \dfrac{x+1}{x+3}$:

$$\int_0^1 \dfrac{1}{(x+3)^2} \ln\left(\dfrac{x+1}{x+3}\right) dx = \dfrac{1}{2}\int_{1/3}^{1/2} \ln y \, dy = \dfrac{1}{2}\Big[y \ln y - y\Big]_{1/3}^{1/2},$$

which gives the required answer. (The integral of $\ln y$ is a standard integral; it can be done by parts, first substituting $z = \ln y$, if you like.)

The second integral can be expressed as the sum of two integrals of the same form as the first integral, since

$$\ln\left(\dfrac{x^2 + 3x + 2}{(x+3)^2}\right) = \ln\left(\dfrac{(x+1)(x+2)}{(x+3)^2}\right) = \ln\left(\dfrac{x+1}{x+3}\right) + \ln\left(\dfrac{x+2}{x+3}\right).$$

We have already done the first of these integrals. Using the substitution $y = \dfrac{x+2}{x+3}$ in the second of these integrals gives

$$\int_{2/3}^{3/4} \ln y \, dy = \dfrac{17}{12}\ln 3 - \dfrac{13}{6}\ln 2 - \dfrac{1}{12}.$$

For the final integral, note that

$$\ln\left(\dfrac{x+1}{x+2}\right) = \ln\left(\dfrac{x+1}{x+3}\right) - \ln\left(\dfrac{x+2}{x+3}\right),$$

so the required integral is the difference of the two integrals that we summed in the previous part, i.e.

$$-\dfrac{5}{4}\ln 3 + \dfrac{23}{12}\ln 2.$$

Post-mortem

I hope you had time to try the original version of this question, given in the comments above. The last part is a nice puzzle.

For the first integral of the original version, the change of variable gives

$$\dfrac{1}{5}\int_{1/3}^{1} y \ln y \, dy = \dfrac{1}{90}\ln 3 - \dfrac{2}{45}.$$

I checked this using Wolfram Alpha. For the second integral, you have to guess the substitution. There are plenty of clues, but the most obvious place to start is the log. The argument of the log can be written as

$$\dfrac{2\sqrt{z} - 5}{\sqrt{z}}.$$

It is now a bit of a leap in the dark (too much of a leap, I thought), but if we decide to convert this exactly into the previous integral, we should take $z = (x+1)^2$. And, magic (or contrived STEP question), it works, giving the first integral almost exactly (note especially the the limits transform as they should). The only difference is a factor of 2 so the answer is twice the previous answer.

A change of variable of the form (∗) is called a *linear fractional* or *Möbius* transformation. It is of great importance in the theory of the geometry of the complex plane.

Problem 33: More curve sketching (✓✓)

The curve C has equation
$$y = \frac{x}{\sqrt{x^2 - 2x + a}},$$
where the square root is positive. Show that, if $a > 1$, then C has exactly one stationary point. Sketch C when (i) $a = 2$ and (ii) $a = 1$.

1999 Paper II

Comments

You have to be sure of the definition of $\sqrt{x^2 - 2x + a}$ to do this question. You might think that it is ambiguous, because the square root could be positive of negative, but by convention it means the *positive* square root. Thus $\sqrt{x^2} = |x|$ (not x).

For the sketches, you just need the position of any stationary points, any other interesting points that the graph passes through, behaviour as $x \to \pm\infty$ and any vertical asymptotes. It should not be necessary in such a simple case to establish the nature of the stationary point(s). The sketches are not particularly difficult, but they do need thought because they turn out to be rather different from the sketches of polynomials or trigonometric functions that you are probably used to.

You should at some stage think about the conditions given on a. In fact, I doubt if many of you will want to leave it there. It is clear that the examiner would really have liked you to do is to sketch C in the different cases that arise according to the value of a but was told that this would be too long and/or difficult for the exam; but I am sure that this is what you will try to do (or maybe what you have already done by the time you read this).

Please do not use your calculator (or Wolfram Alpha or Geogebra or Desmos[24], etc) for these sketches, except perhaps to check your answers: there is no point.

---

[24] https://www.desmos.com/calculator is brilliant: you can use a slider to change the value of a; but please don't.

Solution to problem 33

First differentiate to find the stationary points:

$$\frac{dy}{dx} = \frac{(x^2 - 2x + a) - x(x - 1)}{(x^2 - 2x + a)^{\frac{3}{2}}} = \frac{a - x}{(x^2 - 2x + a)^{\frac{3}{2}}}$$

which is only zero when $x = a$. I've used the quotient rule, but the product rule is just as good.

(i) When $a = 2$, the stationary point is at $(2, \sqrt{2})$ (which can be seen to be a maximum by evaluating the second derivative at $x = a$, though this is not necessary). The curve passes through $(0, 0)$. As $x \to \infty$, $y \to 1$; as $x \to -\infty$, $y \to -1$.

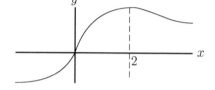

(ii) When $a = 1$, we can rewrite y as $\dfrac{x}{|x - 1|}$. The modulus signs arise because the square root is taken to be positive (or zero — but that can be discounted here since it is in the denominator). The graph is as before, except that the maximum point has been stretched into a vertical asymptote (like a volcanic plug) at $x = 1$. Note that if $|x|$ is much larger than a, it doesn't really matter whether $a = 1$ or $a = 2$.

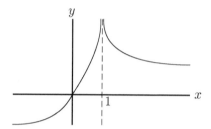

Post-mortem

The first thing to do after finishing the question is to try to understand the role of the parameter a. Part (ii) above is the key. Writing

$$f(x) = x^2 - 2x + a = (x - 1)^2 + a - 1,$$

we see that $f(x) > 0$ for all x if $a > 1$ but if $a < 1$ there are values of x for which $f(x) < 0$. The borderline value is $a = 1$, for which $f(x) \geq 0$ and $f(1) = 0$.

If $a < 1$, we find that $f(x) = 0$ when $x = 1 \pm \sqrt{1 - a}$ and $f(x) < 0$ between these two values of x. The significance of this is that the square root in the denominator of the function we are trying to sketch is imaginary, so there is a gap in the graph between $x = 1 - \sqrt{1 - a}$ and $x = 1 + \sqrt{1 - a}$. There are vertical asymptotes at these values.

Is that the whole story? Well, no. We now have to decide whether the two significant points of our graph, namely $x = 0$ where the curve crosses the axis and $x = a$ where the curve has a maximum, lie inside or outside the forbidden zone $1 - \sqrt{1 - a} \leq x \leq 1 + \sqrt{1 - a}$. This turns out to depend on whether $a < 0$ or $a = 0$ or $0 < a < 1$. The graph will look qualitatively different in each case, so there is still quite a lot of work to do in this $a < 1$ case! We can why it was excluded from the original question.

Problem 34: Trig sum (✓✓✓)

Prove that

$$\sum_{k=0}^{n} \sin k\theta = \frac{\cos \frac{1}{2}\theta - \cos(n + \frac{1}{2})\theta}{2 \sin \frac{1}{2}\theta} . \qquad (*)$$

(i) Deduce that, for $n \geqslant 1$,

$$\sum_{k=0}^{n} \sin\left(\frac{k\pi}{n}\right) = \cot\left(\frac{\pi}{2n}\right) .$$

(ii) By differentiating $(*)$ with respect to θ, or otherwise, show that, for $n \geqslant 1$,

$$\sum_{k=0}^{n} k \sin^2\left(\frac{k\pi}{2n}\right) = \frac{(n+1)^2}{4} + \frac{1}{4}\cot^2\left(\frac{\pi}{2n}\right) .$$

1999 Paper II

Comments

The very first part can be done by multiplying both sides of $(*)$ by $\sin \frac{1}{2}\theta$ and using the identity

$$2 \sin A \sin B = \cos(B - A) - \cos(B + A) .$$

It can also be done by induction (worth a try even if you do it by the above method) or by considering the imaginary part of $\sum \exp(ik\theta)$ (summing this as a geometric progression), if you know about de Moivre's theorem.

Before setting pen to paper for part (ii), it pays to think very hard about simplifications of the right hand side of $(*)$ that might make the differentiation more tractable – perhaps bearing in mind the calculations required for part (i).

In the original question, you were asked to show that, for large n,

$$\sum_{k=0}^{n} \sin\left(\frac{k\pi}{n}\right) \approx \frac{2n}{\pi}$$

and

$$\sum_{k=0}^{n} k \sin^2\left(\frac{k\pi}{2n}\right) \approx \left(\frac{1}{4} + \frac{1}{\pi^2}\right) n^2 ,$$

using the approximations, valid for small θ, $\sin \theta \approx \theta$ and $\cos \theta \approx 1 - \frac{1}{2}\theta^2$. Of course, these results follow quickly from the exact results; but if you only want approximate results you can save yourself a bit of work by approximating early to avoid doing some of the trigonometric calculations. I thought that the exact results were nicer than the approximate results so I changed the question a bit for this book (though I felt a bit guilty, because approximations are an important part of mathematics).

Solution to problem 34

For the first part, we will show that

$$\sum_{k=0}^{n} 2 \sin k\theta \sin \tfrac{1}{2}\theta = \cos \tfrac{1}{2}\theta - \cos(n+\tfrac{1}{2})\theta.$$

Starting with the left hand side, we have

$$\sum_{k=0}^{n} 2 \sin k\theta \sin \tfrac{1}{2}\theta = \sum_{k=0}^{n} \left[-\cos(k+\tfrac{1}{2})\theta + \cos(k-\tfrac{1}{2})\theta \right]$$

which gives the result immediately, since almost all the terms in the sum cancel. (Write our a few terms if you are not certain of this.)

(i) Let $\theta = \pi/n$. Then

$$\sum_{k=0}^{n} \sin\left(\frac{k\pi}{n}\right) = \frac{\cos \tfrac{1}{2}(\pi/n) - \cos(\pi + \tfrac{1}{2}(\pi/n))}{2\sin \tfrac{1}{2}(\tfrac{\pi}{n})} = \frac{\cos \tfrac{1}{2}(\pi/n) + \cos \tfrac{1}{2}(\pi/n)}{2\sin \tfrac{1}{2}(\pi/n)} = \cot \tfrac{1}{2}(\pi/n)$$

as required.

(ii) Differentiating the left hand side of $(*)$ and using a double angle trig. formula gives

$$\sum_{k=0}^{n} k \cos k\theta = \sum_{k=0}^{n} k\left(1 - 2\sin^2 \tfrac{1}{2}k\theta\right) = \tfrac{1}{2}n(n+1) - 2\sum_{k=0}^{n} k\sin^2 \tfrac{1}{2}k\theta. \quad (\dagger)$$

Before attempting to differentiate the right hand side of $(*)$ it is a good idea to write it in a form that gets rid of some of the fractions. Omitting for the moment the factor $1/2$, we have

$$\text{RHS of } (*) = (1 - \cos n\theta) \cot \tfrac{1}{2}\theta + \sin n\theta$$

and differentiating gives $-\tfrac{1}{2}(1 - \cos n\theta)\operatorname{cosec}^2 \tfrac{1}{2}\theta + n\sin n\theta \cot \tfrac{1}{2}\theta + n\cos n\theta$. Now setting $\theta = \pi/n$ gives

$$-\operatorname{cosec}^2 \tfrac{1}{2}(\pi/n) - n. \quad (\ddagger)$$

Setting $\theta = \pi/n$ in equation $(\dagger)$ and comparing with $(\ddagger)$ (remembering that there is a factor of $1/2$ missing) gives the required result.

Post-mortem

Now that I have had another go at this question it does not seem terribly interesting. At first, I thought I might ditch it. Then I thought that it perhaps was worthwhile: keeping cool under the pressure of the differentiation for part (ii) – it should just be a few careful lines – and getting it out is a good confidence booster. Anyway, now that I have slogged[25] through it, I am going to leave it in.

(Next day) I recall now that the reason that I included this question in the first place was because in its original form (with approximate answers as given in the comments section above), the result for part (ii) can be obtained very quickly from part (i) by differentiating both sides with respect to π and using a double angle formula. Try it!

Unfortunately, you'd have a hard time justifying this approach, and anyway it is not going to work for the exact result.[26]

---

[25] I should come clean at this point: I just differentiated $(*)$ on autopilot, without considering whether I could simplify the task by rewriting the formula as suggested in my hint on the previous page. It was hard work. Serves me right for not following my own general advice.

[26] For large n we are just adding up a lot of sin curves and it doesn't much matter what the exact value of π is (so we can take it to be a variable); for the exact result, the exact value of π does matter.

Problem 35: Roots of a cubic equation (✓✓✓)

Consider the cubic equation
$$x^3 - px^2 + qx - r = 0,$$
where $p \neq 0$ and $r \neq 0$.

(i) If the three roots can be written in the form ak^{-1}, a and ak for some constants a and k, show that one root is q/p and that $q^3 - rp^3 = 0$.

(ii) If $r = q^3/p^3$, show that q/p is a root and that the product of the other two roots is $(q/p)^2$. Deduce that the roots are in geometric progression.

(iii) Find a necessary and sufficient condition involving p, q and r for the roots to be in arithmetic progression.

1999 Paper III

Comments

The Fundamental Theorem of Algebra says that a polynomial of degree n can be written as the product of n linear factors, so we can write
$$x^3 - px^2 + qx - r = (x - \alpha)(x - \beta)(x - \gamma), \qquad (*)$$
where α, β and γ are the roots of the equation $x^3 - px^2 + qx - r = 0$. The basis of this question is the comparison between the left hand side of $(*)$ and the right hand side, multiplied out, of $(*)$. Some of the roots may not be real, but you don't have to worry about that here.

The 'necessary and sufficient' in part (iii) looks a bit forbidding, but if you just repeat the steps of parts (i) and (ii), it is straightforward.

Solution to problem 35

(i) We have
$$(x - ak^{-1})(x - a)(x - ak) \equiv x^3 - px^2 + qx - r$$

i.e.
$$x^3 - a(k^{-1} + 1 + k)x^2 + a^2(k^{-1} + 1 + k)x - a^3 = x^3 - px^2 + qx - r \,.$$

Thus $p = a(k^{-1} + 1 + k)$, $q = a^2(k^{-1} + 1 + k)$, and $r = a^3$. Dividing gives $q/p = a$, which is a root, and $q^3/p^3 = a^3 = r$ as required.

(ii) Set $r = q^3/p^3$. Substituting $x = q/p$ into the cubic shows that it is a root:
$$(q/p)^3 - p(q/p)^2 + q(q/p) - (q/p)^3 = 0 \,.$$

Since q/p is one root, and the product of the three roots is q^3/p^3 ($= r$ in the original equation), the product of the other two roots must be q^2/p^2. The two roots can therefore be written in the form $k^{-1}(q/p)$ and $k(q/p)$ for some number k, which shows that the three roots are in geometric progression.

(iii) The three roots are in arithmetic progression if and only if they are of the form $(a - d)$, a and $(a + d)$. (I have followed the lead of part (i) in using this form rather than a, $a + d$, $a + 2d$.)

If the roots are in this form, then
$$\bigl(x - (a - d)\bigr)(x - a)\bigl(x - (a + d)\bigr) \equiv x^3 - px^2 + qx - r$$

i.e.
$$x^3 - 3ax^2 + (3a^2 - d^2)x - a(a^2 - d^2) \equiv x^3 - px^2 + qx - r \,.$$

Thus $p = 3a$, $q = 3a^2 - d^2$, $r = a(a^2 - d^2)$. A necessary condition is therefore $r = (p/3)(q - 2p^2/9)$. Note that one root is $p/3$.

Conversely, if $r = (p/3)(q - 2p^2/9)$, the equation is
$$x^3 - px^2 + qx - (p/3)(q - 2p^2/9) = 0 \,.$$

We can verify that $p/3$ is a root by substitution. Since the roots sum to p and one of them is $p/3$, the others must be of the form $p/3 - d$ and $p/3 + d$ for some d. They are therefore in arithmetic progression.

A necessary and sufficient condition for the roots to be in arithmetic progression is therefore $r = (p/3)(q - 2p^2/9)$.

Post-mortem

As usual, it is a good idea to give a bit of thought to the conditions given, namely $p \ne 0$ and $r \ne 0$.

Clearly, if the roots are in geometric progression, we cannot have a zero root. We therefore need $r \ne 0$. However, we don't need the condition $p \ne 0$. If the roots are in geometric progression with $p = 0$, then $q = 0$, but there is no contradiction: the roots are $be^{-2\pi i/3}$, b and $be^{2\pi i/3}$ where $b^3 = r$, and these are certainly in geometric progression. So the necessary and sufficient conditions are $r \ne 0$ and $p^3 r = q^3$.

Neither of these conditions is required if the roots are in arithmetic progression.

Therefore, the condition $p \ne 0$ is only there as a convenience — one thing less for you to worry about. The condition $r \ne 0$ should really have been given only for the first part. It should be said that this sort of question is incredibly difficult to word, which is why the examiner was a bit heavy handed with the conditions.

Problem 36: Root counting (✓✓✓)

(i) Let $f(x) = (1+x^2)e^x - k$, where k is a given constant. Show that $f'(x) \geqslant 0$ and sketch the graph of $f(x)$ in the three cases $k < 0$, $0 < k < 1$ and $k > 1$.

Hence, or otherwise, show that the equation

$$(1+x^2)e^x - k = 0$$

has exactly one real root if $k > 0$ and no real roots if $k \leqslant 0$.

(ii) Determine the number of real roots of the equation

$$(e^x - 1) - k \tan^{-1} x = 0$$

in the different cases that arise according to the value of the constant k.

Note: If $y = \tan^{-1} x$, then $\dfrac{dy}{dx} = \dfrac{1}{1+x^2}$.

1999 Paper III

Comments

In good STEP style, this question has two related parts. In this case, the first part not only gives you guidance for the second part, but also provides a result that is useful for the second part.

In part (ii) of the original question you were asked to determine the number of real roots in the cases $0 < k \leqslant 2/\pi$ and $2/\pi < k < 1$. I thought that you would like to work out all the different cases for yourself — but it makes the question considerably harder, especially if you think about the special values of k as well as the ranges of k.

For part (i) you need to know that $xe^x \to 0$ as $x \to -\infty$. This is just a special case of the result that exponentials go to zero faster than any power of x.

For part (ii), it is not really necessary to do all the sketches, but I think you should (though you perhaps wouldn't under examination conditions) because it gives you a complete understanding of the way the function depends on the parameter k. For the graphs, you may want to consider the sign of $f'(0)$; this will give you an important clue. Remember that, by definition, $-\tfrac{1}{2}\pi < \tan^{-1} x < \tfrac{1}{2}\pi$.

Solution to problem 36

(i) Differentiating gives $f'(x) = 2xe^x + (1+x^2)e^x = (1+x)^2 e^x$ which is non-negative because the square is non-negative and the exponential is positive.

There is one stationary point, at $x = -1$, which is a point of inflection, since the gradient on either side is positive.

Also, $f(x) \to -k$ as $x \to -\infty$, $f(0) = 1 - k$ and $f(x) \to \infty$ as $x \to +\infty$. The graph is essentially exponential, with a hiccup at $x = -1$. The differences between the three cases are the positions of the horizontal asymptote ($x \to -\infty$) and the place where the graph cuts the y axis (above or below the x axis?).

Since the graph is increasing and $0 < f(x) < \infty$, the given equation has one real root if $k > 0$ and no real roots if $k \leqslant 0$.

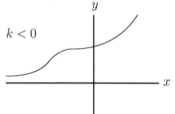
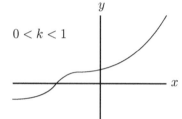
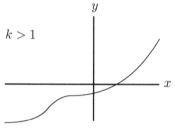

(ii) First we collect up information required to sketch the graph, as in part (i). We have
$$f(0) = 0, \quad f(x) \to \tfrac{1}{2}k\pi - 1 \text{ as } x \to -\infty, \quad f(x) \to \infty \text{ as } x \to \infty, \quad f'(x) = e^x - k(1+x^2)^{-1}$$ (which we know from the first part can only be zero only if $k > 0$), $f'(0) = 1 - k$.

The important values of k seem to be 0, $2/\pi$ and 1 so we will have to consider four cases.

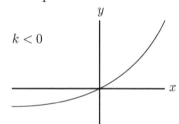
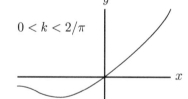
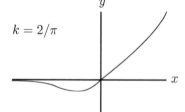

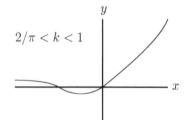
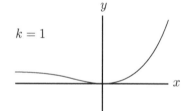
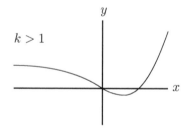

As you see, the equation $(e^x - 1) - k \tan^{-1} x = 0$ has one root if $k \leqslant \tfrac{1}{2}\pi$ and two roots otherwise. The case $k = 1$ can be thought of as having one root at $x = 0$ or as having two roots, both at $x = 0$ (it is a double root).

Post-mortem

I think I made a bit of a meal of this; got carried away with the sketches. But it was reassuring (compulsive, I found) to consider every case. My explanation is pretty compact, so I think you'll have to work it through on your own to convince yourself of the details.

Problem 37: Irrationality of e (✓✓)

For each positive integer n, let

$$a_n = \frac{1}{n+1} + \frac{1}{(n+1)(n+2)} + \frac{1}{(n+1)(n+2)(n+3)} + \cdots,$$

$$b_n = \frac{1}{n+1} + \frac{1}{(n+1)^2} + \frac{1}{(n+1)^3} + \cdots.$$

(i) Show that $b_n = 1/n$.

(ii) Deduce that $0 < a_n < 1/n$.

(iii) Show that $a_n = n!\,e - [n!\,e]$, where $[x]$ denotes the integer part of x.

(iv) Hence show that e is irrational.

1997 Paper III

Comments

Each part of this looks horrendously difficult, but it doesn't take much thought to see what is going on. If you haven't come across the concept of *integer part* some examples should make it clear: $[21.25] = 21$, $[\pi] = 3$, $[1.999999] = 1$ and $[2] = 2$.

For part (iii), you need to know the series for the number e, which is $\sum_{n=0}^{\infty} 1/n!$.

If you are stumped by the last part, just remember the definition of the word *irrational*: x is rational if and only if it can be expressed in the form p/q where p and q are integers; if x is not rational, it is irrational. Then look for a proof by contradiction ('Suppose that x is rational ...').

The first proof that e is irrational was first given by Euler (1736). He also named the number e, though he didn't 'invent' it; e is the basis for natural logarithms and as such was used implicitly by John Napier in 1614. Euler's proof involved the use of continuous fractions. He wrote e as

$$2 + \cfrac{1}{1 + \cfrac{1}{1 + \cfrac{1}{4 + \cfrac{1}{\cdots}}}}$$

the numbers in the denominators forming the infinite pattern $1, 2, 1, 1, 4, 1, 1, 6, 1, 1, 8 \ldots$. It is known that any such continuous fraction represents an irrational number.

In general, it is not easy to show that numbers are irrational. Johann Lambert showed that π is irrational in 1760, but the simple continued fraction for π is not known. It is still not known if $\pi + e$ is irrational.

The proof that e is irrational given in this question is based on a proof by Fourier[27] in about 1815.

---

[27] Joseph Fourier, 1768–1830, was a French mathematician and administrator. He is best known for his work on the theory of heat flow. He is generally credited with the discovery of the greenhouse effect. He spent some time in Egypt with Napoleon, and contributed much to the modern study of Egyptology.

Solution to problem 37

(i) Although the series for b_n does not at first sight look tractable, it is in fact just a geometric progression: the first term is $1/(n+1)$ and the common ratio is also $1/(n+1)$. Thus

$$b_n = \frac{1}{n+1}\left(\frac{1}{1 - 1/(n+1)}\right) = \frac{1}{n}.$$

(ii) Each term (after the first) of a_n is less than the corresponding term in b_n, so $a_n < b_n = 1/n$.

(iii) Multiplying the series for e by $n!$ gives

$$n!e = n! + n! + \tfrac{1}{2}n! + \cdots + 1 + a_n$$

and the result follows because $a_n < 1$ and all the other terms on the right hand side of the above equation are integers.

(iv) We use proof by contradiction. Suppose that there exist integers k and m such that $e = k/m$. Then $m!e$ is certainly an integer. But if $m!e$ is an integer then $[m!e] = m!e$, which contradicts the result of part (iii) since we know that $a_m \neq 0$ (it is obvious from the definition that $a_m > 0$).

Post-mortem

Well, that was short!

There were, however, two key steps which were not difficult in themselves but were not easy to find in the context of this question.

The first was to recognise that b_n is, for each n, a simple geometric progression. If it had been presented in the form $b = r + r^2 + r^3 + \cdots$ it would have been immediately recognisable. Somehow, the fact that the ratio r is given in terms of n — or, even worse, $n+1$ — makes it difficult to spot; not least because n is fixed for each series instead of labelling the terms of an individual series.

The second key step, in part (iv), was to assume that $e = k/m$ and show that $m!e$, **not** me, cannot be an integer. It seems so wasteful when it would suffice to show that me is not an integer. We seem to be making a great deal of extra work for ourselves and that profligacy, unusual in mathematics, might well have thrown you off the scent despite the clear signal from part (iii).

I don't regard the idea of using proof by contradiction as a difficult step in part (iv). An irrational number is by definition a number that is not rational, so the idea of a proof by contradiction should jump out at you. Indeed, the standard example, the proof of the irrationality of $\sqrt{2}$, is by contradiction.[28]

---

[28] Suppose $p\sqrt{2} = q$. Then $2p^2 = q^2$. Writing p and q as the product of primes gives the contradiction since there are an odd number of powers of 2 one side of this equation and an even number on the other.

Problem 38: Discontinuous integrands (✓✓)

For any number x, the largest integer less than or equal to x is denoted by $[x]$. For example, $[3.7] = 3$ and $[4] = 4$.

(i) Sketch the graph of $y = [x]$ for $0 \leqslant x < 5$ and evaluate

$$\int_0^5 [x]\, dx .$$

(ii) Sketch the graph of $y = [e^x]$ for $0 \leqslant x < \ln n$, where n is an integer, and show that

$$\int_0^{\ln n} [e^x]\, dx = n \ln n - \ln(n!) .$$

Hence show that $n! \geqslant n^n e^{1-n}$.

Comments

I had to add a bit to the original question because it was all dressed up with nowhere to go. The question is clearly about estimating $n!$ so I added in the last line, which makes the question a bit longer but not much more difficult. I could have added another part, but I thought that you would probably add it yourself: you will easily spot, once you have drawn the graphs, that a similar result for $n!$ with the inequality reversed can be obtained by considering rectangles the tops of which are above the graph of e^x instead of below it. You therefore end up with a nice sandwich inequality for $n!$.

Stirling (1692 – 1770) proved in 1730 that $n! \approx \sqrt{2\pi}\, n^{n+\frac{1}{2}} e^{-n}$ for large n. This was a brilliant result — even reading his book, it is hard to see where he got $\sqrt{2\pi}$ from (especially as he wrote in Latin). Then he went on to obtain the approximation in terms of an infinite series; the expression above is just the first term.

Solution to problem 38

(i)

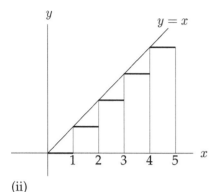

The graph of $[x]$ consists of the horizontal parts of an ascending staircase with 5 stairs, the lowest at height 0, each of width 1 unit and rising 1 unit.

The integral is the sum of the areas of the rectangles shown in the figure:
$$\text{area} = 0 + 1 + 2 + 3 + 4 = 10.$$

(ii)

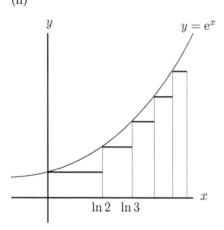

The graph of $[e^x]$ is also a staircase: the height of each stair is 1 unit and the width decreases as x increases because the gradient of e^x increases. It starts at height 1 and ends at height $(n-1)$.

The value of $[e^x]$ changes from $k-1$ to k when $e^x = k$ i.e. when $x = \ln k$. Thus $[e^x] = k$ when $\ln k \leqslant x < \ln(k+1)$ and the area of the corresponding rectangle is $k\big(\ln(k+1) - \ln k\big)$. The total area under the curve is therefore

$$1(\ln 2 - \ln 1) + 2(\ln 3 - \ln 2) + \cdots + (n-1)\big(\ln n - \ln(n-1)\big)$$

which you can rearrange to obtain $n \ln n - \ln(n!)$ as required.

For the last part, note that $[e^x] \leqslant e^x$ (this is clear from the definition) so

$$\int_0^{\ln n} [e^x]\, dx \leqslant \int_0^{\ln n} e^x\, dx = n - 1 \quad \text{i.e.} \quad n \ln n - \ln(n!) \leqslant n - 1.$$

Taking exponentials gives the required result.

Post-mortem

Usually when you draw a graph of a discontinuous function, you should specify at the jump whether the function takes the upper or lower value. For example $[x]$ takes the value 1 at $x = 1$, which is the upper value. This can be achieved by putting (say) a circle round the upper or lower point, as appropriate. I didn't bother on the above graphs, because it doesn't affect the value of the integral and I didn't want to clutter up the graphs.

Considering $[e^x + 1]$ instead of $[e^x]$ gives rectangles above the graph of $y = e^x$ rather than below. The calculations are roughly the same, so you should easily arrive at $n! \leqslant n^{n+1} e^{1-n}$. We have therefore proved that

$$n^n e^{1-n} \leqslant n! \leqslant n^{n+1} e^{1-n}.$$

This gives a pretty good (given the rather elementary method at our disposal) approximation for $n!$ for large n.

Problem 39: A difficult integral (✓✓✓)

Given that $\tan \frac{1}{4}\pi = 1$ show that $\tan \frac{1}{8}\pi = \sqrt{2} - 1$.

Let
$$I = \int_{-1}^{1} \frac{1}{\sqrt{1+x} + \sqrt{1-x} + 2} \, dx \, .$$

Show, by using the change of variable $x = \sin 4t$, that

$$I = \int_{0}^{\frac{1}{8}\pi} \frac{2\cos 4t}{\cos^2 t} \, dt \, .$$

Hence show that

$$I = 4\sqrt{2} - \pi - 2 \, .$$

1996 Paper II

Comments

This tests trigonometric manipulation and integration skills. You will certainly need $\tan 2\theta$ in terms of $\tan \theta$, and $\cos 2\theta$ in terms of $\cos \theta$, and maybe other formulae.

Both parts of the question are *multistep*: there are half a dozen consecutive steps, each different in nature, with no guidance. This is unusual in school-level mathematics but normal in university mathematics.

I checked the answer on Wolfram Alpha, which turned out to be very good indeed at doing this sort of thing. I asked it to do the indefinite integral as well and, in less than a second, it came up with

$$\sqrt{1-x}\left[-1 - (\sqrt{x+1} + 1)^{-1}\right] + [\sqrt{x+1} + 1]^{-1} - 2\arcsin\sqrt{\tfrac{1}{2}(x+1)} \, .$$

Not a pretty sight and not in its neatest form by a long way: for example, $2\arcsin\sqrt{\tfrac{1}{2}(x+1)}$ reduces, after a bit of algebra, to $\tfrac{1}{2}\pi + \arcsin x$. I also asked it to do the same integral with the 2 replaced by a parameter k and it took four seconds. The answer was about 20 times longer than the $k = 2$ result but it seemed to enjoy the problem, as far as I could tell.

Solution to problem 39

For the first part, to save writing, let $t = \tan\frac{\pi}{8}$. Then

$$\frac{2t}{1-t^2} = 1 \Rightarrow t^2 + 2t - 1 = 0 \Rightarrow t = \frac{-2 \pm \sqrt{8}}{2} = -1 \pm \sqrt{2}\,.$$

We take the root with the $+$ sign since we know that t is positive.

Now the integral. Note first that the integrand has an obvious symmetry: it is unchanged when $x \leftrightarrow -x$. This means that we can do the integral over the half-range $x = 0$ to $x = 1$ and double the result. A glance at the required result suggests that this is a good idea.

Substituting $x = \sin 4t$ as instructed then gives

$$I = 2 \int_0^{\frac{1}{8}\pi} \frac{4\cos 4t}{\sqrt{1+\sin 4t} + \sqrt{1-\sin 4t} + 2}\, dt$$

so to obtain the given answer we need to show that

$$\sqrt{1+\sin 4t} + \sqrt{1-\sin 4t} + 2 = 4\cos^2 t\,,$$

i.e.

$$\sqrt{1+\sin 4t} + \sqrt{1-\sin 4t} = 2\cos 2t\,.$$

If we square both sides of this equation, noting that both sides are positive for the values of t in the integral so this is not dangerous, nice things happen:

$$(1+\sin 4t) + (1-\sin 4t) + 2\sqrt{1-\sin^2 4t} = 4\cos^2 2t \qquad \text{(RTP)}$$

i.e.

$$2 + 2\cos 4t = 4\cos^2 2t \qquad \text{(RTP)}$$

which is true by a standard trig. identity, so we have proved what we were required to prove.

For the last part, we have

$$\cos 4t = 2\cos^2 2t - 1 = 2(2\cos^2 t - 1)^2 - 1 = 8\cos^4 t - 8\cos^2 t + 1$$

so

$$I = 2\int_0^{\frac{1}{8}\pi} (8\cos^2 t - 8 + \sec^2 t)\, dt = 2\int_0^{\frac{1}{8}\pi} (4\cos 2t - 4 + \sec^2 t)\, dt = 2\Big[2\sin 2t - 4t + \tan t\Big]_0^{\frac{1}{8}\pi},$$

which gives the required result.

Post-mortem

Manipulating the integrand after the change of variable was really quite demanding. You could easily go down the wrong track and become mired in algebra. I did it by writing down what I was trying to prove and then showing that it was indeed true. This of course is hazardous, because if you are not careful you might assume the result in order to prove the result. I find it helps to write 'RTP' (Required To Prove) in the margin to indicate clearly to myself and others that I am not assuming it to be true.

The first thing we did with the integral, guided by the given answer, was to use the symmetry $x \leftrightarrow -x$ to reduce the range of integration to $0 \leqslant x \leqslant 1$, doubling the result. It is clear from a graph that this works but you could, if you were unsure, split the integral into two parts (integral from -1 to 0 plus integral from 0 to 1) then make the change of variable $y = -x$ in the lower integral to show that the two parts are equal.

Problem 40: Estimating the value of an integral (✓✓)

(i) Show that, for $0 \leqslant x \leqslant 1$, the largest value of $\dfrac{x^6}{(x^2+1)^4}$ is $\dfrac{1}{16}$.

What is the smallest value?

(ii) Find constants A, B, C and D such that, for all x,

$$\frac{1}{(x^2+1)^4} \equiv \frac{d}{dx}\left(\frac{Ax^5 + Bx^3 + Cx}{(x^2+1)^3}\right) + \frac{Dx^6}{(x^2+1)^4}.$$

(iii) Hence, or otherwise, prove that

$$\frac{11}{24} \leqslant \int_0^1 \frac{1}{(x^2+1)^4}\, dx \leqslant \frac{11}{24} + \frac{1}{16}.$$

2000 Paper I

Comments

You should think about part (i) graphically, though it is not necessary to draw the graph: just set about it as if you were going to (starting point, finishing point, turning points, etc).

The equivalence sign in part (ii) indicates an equality that holds for all x — you are not being asked to solve the equation for x.

For part (iii), you need to know that inequalities can be integrated: this is 'obvious' if you think about integration in terms of area, though a formal proof requires a formal definition of integration and this is the sort of thing you would do in a first university course in mathematical analysis.

Solution to problem 40

(i) Let $f(x) = \dfrac{x^6}{(x^2+1)^4}$. Then

$$f'(x) = \frac{6x^5}{(x^2+1)^4} - \frac{8x^7}{(x^2+1)^5} = \frac{2(3-x^2)x^5}{(x^2+1)^5}$$

which is positive for $0 < x^2 < 3$. Therefore, $f(x)$ increases in value from 0 at $x=0$ to $\frac{1}{16}$ at $x=1$.

(ii) Doing the differentiation gives

$$\frac{1}{(x^2+1)^4} \equiv \frac{5Ax^4 + 3Bx^2 + C}{(x^2+1)^3} - \frac{6x(Ax^5 + Bx^3 + Cx)}{(x^2+1)^4} + \frac{Dx^6}{(x^2+1)^4}$$

and multiplying by $(x^2+1)^4$ gives the following identity:

$$1 \equiv (5Ax^4 + 3Bx^2 + C)(x^2+1) - 6x(Ax^5 + Bx^3 + Cx) + Dx^6 \qquad (*)$$

i.e.

$$1 \equiv (D-A)x^6 + (5A-3B)x^4 + (3B-5C)x^2 + C\,.$$

Equating coefficients of the different powers of x on each side of the equivalence sign gives $1 = C$, $0 = 3B - 5C$, $0 = 5A - 3B$, $0 = D - A$, so $A = 1$, $B = \frac{5}{3}$, $C = 1$ and $D = 1$.

(iii) Using the results of parts (i) and (ii), we see that for $0 \leqslant x \leqslant 1$

$$\frac{d}{dx}\left(\frac{x^5 + \frac{5}{3}x^3 + x}{(x^2+1)^3}\right) \leqslant \frac{1}{(x^2+1)^4} \leqslant \frac{d}{dx}\left(\frac{x^5 + \frac{5}{3}x^3 + x}{(x^2+1)^3}\right) + \frac{1}{16}\,.$$

Inequalities can be integrated, so

$$\int_0^1 \frac{d}{dx}\left(\frac{x^5 + \frac{5}{3}x^3 + x}{(x^2+1)^3}\right)dx \leqslant \int_0^1 \frac{1}{(x^2+1)^4}\,dx \leqslant \int_0^1 \frac{d}{dx}\left(\frac{x^5 + \frac{5}{3}x^3 + x}{(x^2+1)^3}\right)dx + \int_0^1 \frac{1}{16}\,dx$$

i.e.

$$\left[\frac{x^5 + (5/3)x^3 + x}{(x^2+1)^3}\right]_0^1 \leqslant \int_0^1 \frac{1}{(x^2+1)^4}\,dx \leqslant \left[\frac{x^5 + (5/3)x^3 + x}{(x^2+1)^3}\right]_0^1 + \left[\frac{x}{16}\right]_0^1$$

from which the required result follows immediately.

Post-mortem

Instead of equating coefficients in $(*)$, you could obtain equations for A, B, C and D by putting four carefully chosen values of x into the equation. An obvious choice is $x = 0$, but (thinking flexibly!) you could try $x = i$ to eliminate terms with factors of $x^2 + 1$. Then it becomes more difficult to find good choices.

You might wonder why, in part (ii), the term inside the derivative has only odd powers of x. Would it not make it more general to include even powers as well? You could in fact include even powers but you would find that their coefficients would be zero: all the other terms in the equation are even in x, so the derivative has to be an even function which means that the function being differentiated must be odd.

Although the idea of this question is good, the final result is a bit feeble. It only gives the value of the integral to an accuracy of about $\frac{1}{16}/\frac{11}{24}$ which is about 15%. The actual value of the integral can be found fairly easily using the substitution $x = \tan t$ and is $\frac{11}{48} + \frac{5}{64}\pi$ so the inequalities can be used to give (rather bad) estimates for π namely $2.933 \leqslant \pi \leqslant 3.733$.

Problem 41: Integrating the modulus function (✓✓)

Show that
$$\int_{-1}^{1} |xe^x|\,dx = -\int_{-1}^{0} xe^x\,dx + \int_{0}^{1} xe^x\,dx$$
and hence evaluate the integral.

Evaluate the following integrals:

(i) $\displaystyle\int_{0}^{4} |x^3 - 2x^2 - x + 2|\,dx$;

(ii) $\displaystyle\int_{-\pi}^{\pi} |\sin x + \cos x|\,dx$.

2000 Paper I

Comments

The very first part shows you how to do this sort of integral (with mod signs in the integrand) by splitting up the range of integration at the points where the integrand changes sign. In parts (i) and (ii) you have to use the technique on different examples.

Solution to problem 41

For the first part, note that $|xe^x| = -xe^x$ if $x < 0$. Then integrate by parts to evaluate the integrals:

$$\int_{-1}^0 (-xe^x)\,dx + \int_0^1 xe^x\,dx = -\left(\Big[xe^x\Big]_{-1}^0 - \int_{-1}^0 e^x\,dx\right) + \left(\Big[xe^x\Big]_0^1 - \int_0^1 e^x\,dx\right) = 2 - 2e^{-1}.$$

For the next parts, we have to find out where the integrand is positive and where it is negative.

(i) $x^3 - 2x^2 - x + 2 = (x-1)(x+1)(x-2)$ (spotting the factors), so the integrand is positive for $0 \leqslant x < 1$, negative for $1 < x < 2$ and positive for $2 < x < 4$ (a quick sketch will help with this). Splitting the range of integration into these ranges and integrating gives

$$\int_0^4 |x^3 - 2x^2 - x + 2|\,dx \tag{*}$$

$$= \int_0^1 (x^3 - 2x^2 - x + 2)\,dx - \int_1^2 (x^3 - 2x^2 - x + 2)\,dx + \int_2^4 (x^3 - 2x^2 - x + 2)\,dx$$

$$= 22\tfrac{1}{6}.$$

(ii) $\sin x + \cos x$ changes sign when $\tan x = -1$, i.e. when $x = -\tfrac{1}{4}\pi$ and $x = \tfrac{3}{4}\pi$. Splitting the range of integration into these ranges and integrating gives

$$\int_{-\pi}^{\pi} |\sin x + \cos x|\,dx = \int_{-\pi}^{-\frac{1}{4}\pi} -(\sin x + \cos x)\,dx + \int_{-\pi/4}^{\frac{3}{4}\pi} (\sin x + \cos x)\,dx + \int_{\frac{3}{4}\pi}^{\pi} -(\sin x + \cos x)\,dx$$

$$= \Big[\cos x - \sin x\Big]_{-\pi}^{-\frac{1}{4}\pi} + \Big[-\cos x + \sin x\Big]_{-\frac{1}{4}\pi}^{\frac{3}{4}\pi} + \Big[+\cos x - \sin x\Big]_{\frac{3}{4}\pi}^{\pi}$$

$$= 4\sqrt{2}.$$

Alternatively, start by writing $\sin x + \cos x = \sqrt{2}\cos(x - \tfrac{1}{4}\pi)$ which makes the changes of sign easier to spot and the integrals easier to do.

Post-mortem

If you give a bit more thought to part (ii), you will see easier ways of doing it. Since the trigonometric functions are periodic with period 2π the integrand is also periodic. Write the integrand in the form $\sqrt{2}|\cos(x - \tfrac{1}{4}\pi)|$. Integrating this over any 2π interval gives the same result. Indeed, we may as well integrate $\sqrt{2}|\sin x|$ from 0 to 2π; or from 0 to π and double the answer.

After the first edition of this book appeared, a correspondent e-mailed to suggest that we should integrate (*) by writing it in the form

$$\int_0^4 (x^3 - 2x^2 - x + 2)\,dx - 2\int_1^2 (x^3 - 2x^2 - x + 2)\,dx.$$

Yes, that's quite a good idea; it would have saved a bit of writing and reduced the risk of arithmetical errors.

Problem 42: Geometry (✓✓)

> In the triangle ABC, angle $BAC = \alpha$ and angle $CBA = 2\alpha$, where 2α is acute, and $BC = x$. Show that $AB = (3 - 4\sin^2 \alpha)x$.
>
> The point D is the midpoint of AB and the point E is the foot of the perpendicular from C to AB. Find an expression for DE in terms of x.
>
> The point F lies on the perpendicular bisector of AB and is a distance x from C. The points F and B lie on the same side of the line through A and C. Show that the line FC trisects the angle ACB.

2015 Paper II

Comments

I thought it would be good to include at least one plane geometry question in this collection, so here it is. As always with geometry, the first thing to do is draw a BIG diagram. You will probably need to have quite a few tries at it.

Then you have the usual tools at your disposal: similar triangles, congruent triangles, angle-chasing, and — moving away from classical Euclidean geometry — Pythagoras, and sine and cosine rules. I leave it to you to decide what will be useful here.

Solution to problem 42

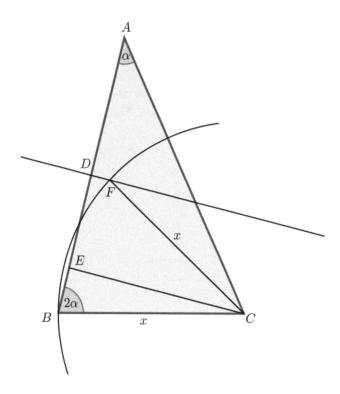

Here is a careful diagram, with a construction arc left in to show that $CB = CF = x$. It took me ages. But it is pretty much plain sailing now.

By the sine rule in $\triangle ABC$ (and calculating an expression for $\sin 3\alpha$ using double angle formulae)

$$AB = \frac{x\sin(180° - 3\alpha)}{\sin\alpha} = \frac{x\sin(3\alpha)}{\sin\alpha} = \frac{x(3\cos^2\alpha\sin\alpha - \sin^3\alpha)}{\sin\alpha} = (3 - 4\sin^2\alpha)x\,.$$

as required.

Then, since D is the mid-point of AB and $\angle BEC = 90°$,

$$DE = \tfrac{1}{2}AB - BE = \tfrac{1}{2}x(3 - 4\sin^2\alpha) - x\cos 2\alpha = \tfrac{1}{2}x,$$

rather surprisingly.

For the last part, since $\angle BCA = 180° - 3\alpha$, we need to show that $\angle FCA = 60° - \alpha$ or, equivalently, $\angle FCB = 120° - 2\alpha$. Now $\angle BCE = 90° - 2\alpha$ so we are done if we can show that $\angle ECF = 30°$. But that follows almost immediately from the $DE = \tfrac{1}{2}x$ and $FC = x$: draw the perpendicular from F to CE, which has length $\tfrac{1}{2}x$, and consider the right-angled triangle thus formed.

Post-mortem

I liked the mixture here of angle-chasing and use of the sine rule. Geometry purists, though, would be appalled; using the sine rule in a geometry problem is like using metal screws instead of wooden dowels on an antique wooden cabinet. As it happens, there is a beautiful proof of this result by pure geometry. You have to reflect the whole diagram about the perpendicular bisector of AB. For the details, see Ross Honsberger's excellent *Mathematical Chestnuts from around the World* (Cambridge University Press, 2001), section 20.

Problem 43: The t substitution (✓✓)

Show that

$$\sin\theta = \frac{2t}{1+t^2}, \quad \cos\theta = \frac{1-t^2}{1+t^2}, \quad \text{and} \quad \frac{1+\cos\theta}{\sin\theta} = \tan(\tfrac{1}{2}\pi - \tfrac{1}{2}\theta),$$

where $t = \tan(\tfrac{1}{2}\theta)$.

Let

$$I = \int_0^{\frac{1}{2}\pi} \frac{1}{1+\cos\alpha\sin\theta}\,d\theta.$$

Use the substitution $t = \tan(\tfrac{1}{2}\theta)$ to show that, for $0 < \alpha < \tfrac{1}{2}\pi$,

$$I = \int_0^1 \frac{2}{(t+\cos\alpha)^2 + \sin^2\alpha}\,dt.$$

By means of the further substitution $t + \cos\alpha = \sin\alpha\tan u$ show that

$$I = \frac{\alpha}{\sin\alpha}.$$

Deduce a similar result for

$$\int_0^{\frac{1}{2}\pi} \frac{1}{1+\sin\alpha\cos\phi}\,d\phi.$$

2000 Paper II

Comments

The first of the two substitutions is familiarly known as the 't substitution'. It would have been very standard fare 30 years ago, but it seems to have gone out of fashion now. The second of the two substitutions is the normal substitution for integrals with quadratic denominators.

For the last part, 'deduce' implies that you don't have to do any further integration. Note that the variable in the integral is ϕ instead of θ. Since it is a definite integral, it doesn't matter what the variable is called; it could equally well have been called θ as in the original integral I. The use of a different variable was just a kindness on the part of the examiner to indicate that you should be thinking about a change of variable.

The 'similar result' result that you deduce should include the conditions under which it is true. It is worth thinking about why the condition $0 < \alpha < \tfrac{1}{2}\pi$ is required — or, indeed, if it is required.

Solution to problem 43

The three identities just require use of $\cos\theta = \cos^2\frac{1}{2}\theta - \sin^2\frac{1}{2}\theta$ and $\sin\theta = 2\sin\frac{1}{2}\theta\cos\frac{1}{2}\theta$. If you divide each by $\cos^2\frac{1}{2}\theta + \sin^2\frac{1}{2}\theta$ (i.e. by 1) the first two identities drop out. Remember, for the last one, that $\cot x = \tan(\frac{1}{2}\pi - x)$.

For the first change of variable, we have $dt = \frac{1}{2}\sec^2\frac{1}{2}\theta\, d\theta = \frac{1}{2}(1+t^2)d\theta$ and the new limits are 0 and 1, so

$$I = \int_0^{\frac{1}{2}\pi} \frac{1}{1+\cos\alpha\sin\theta}\, d\theta = \int_0^1 \frac{1}{1+\cos\alpha\frac{2t}{1+t^2}}\frac{2}{1+t^2}\, dt$$

$$= \int_0^1 \frac{2}{1+2t\cos\alpha + t^2}\, dt = \int_0^1 \frac{2}{(t+\cos\alpha)^2 + \sin^2\alpha}\, dt\, .$$

For the second change of variable, we have $dt = \sin\alpha\sec^2 u\, du$. When $t = 0$, $\tan u = \cot\alpha$ so $u = \frac{1}{2}\pi - \alpha$. When $t = 1$, $\sin\alpha\tan u = 1 + \cos\alpha$ so (after a bit of work with double-angle formulae) $u = \frac{1}{2}\pi - \frac{1}{2}\alpha$. Thus

$$I = \int_{\frac{1}{2}\pi-\alpha}^{\frac{1}{2}\pi-\frac{1}{2}\alpha} \frac{2}{\sin^2\alpha(1+\tan^2 u)}\sin\alpha\sec^2 u\, du = \int_{\frac{1}{2}\pi-\alpha}^{\frac{1}{2}\pi-\frac{1}{2}\alpha} \frac{2}{\sin\alpha}\, du = \frac{\alpha}{\sin\alpha}\, .$$

For the last part, we want to make a substitution that changes the cosine in the denominator to a sine. One possibility is to set $\phi = \frac{1}{2}\pi - \theta$. This will swap the limits but also introduces a minus sign since $d\theta = -d\phi$. Thus

$$\frac{\alpha}{\sin\alpha} = I = -\int_{\frac{\pi}{2}}^0 \frac{1}{1+\cos\alpha\cos\phi}\, d\phi = \int_0^{\frac{\pi}{2}} \frac{1}{1+\cos\alpha\cos\phi}\, d\phi\, .$$

This is almost the integral we want: we still need to replace $\cos\alpha$ in the denominator by $\sin\alpha$. Remembering what we did a couple of lines back, we just replace α by $\frac{1}{2}\pi - \alpha$ in the integral and in the answer, giving

$$\int_0^{\frac{\pi}{2}} \frac{1}{1+\cos\alpha\cos\phi}\, d\phi = \frac{\frac{1}{2}\pi - \alpha}{\cos\alpha}\, .$$

If the original α satisfied $0 < \alpha < \frac{1}{2}\pi$, the new α must satisfy $\frac{1}{2}\pi > \alpha > 0$, which is the same.

Post-mortem

One reason for the restriction on α might be to prevent the denominator of the integrand being zero for some value of θ; a zero in the denominator usually means that the integral is undefined. However, the only value of α for which $\cos\alpha\sin\theta$ could possibly be as small as -1 (for $0 \leq \theta \leq \frac{1}{2}\pi$) is π (and of course 3π, etc). From this point of view, we only need $\alpha \neq \pi$ (etc). There is a slight awkwardness in the answer when $\alpha = 0$ but this can be overcome by taking limits:

$$\lim_{\alpha \to 0} \frac{\alpha}{\sin\alpha} = 1$$

which you can easily verify is the correct value of the integral when $\alpha = 0$.

You might therefore think that the restriction $0 < \alpha < \frac{1}{2}\pi$ is superfluous. But here is a curious thing: increasing α by 2π does not change I but does change the answer! If you work through the solution with this in mind you see that it can make a difference only when you work out $\tan^{-1} u$, which by definition lies in the range $-\frac{1}{2}\pi$ to $\frac{1}{2}\pi$, and this is why we must have $-\frac{1}{2}\pi < \alpha < \frac{1}{2}\pi$. We lose nothing by using instead $0 \leq \alpha \leq \frac{1}{2}\pi$, since $\alpha \to -\alpha$ doesn't change the integral. The strict inequalities ($<$ rather than $\leq$) avoid trouble with the denominators when $\alpha = 0$ in the first integral or $\alpha = \frac{1}{2}\pi$ in the second.

Problem 44: A differential-difference equation (✓✓)

A damped system with feedback is modelled by the equation

$$f'(t) = -f(t) + kf(t-1), \tag{1}$$

where k is a given non-zero constant.

Show that (non-zero) solutions for f of the form $f(t) = Ae^{mt}$, where A and m are constants, are possible provided m satisfies

$$m + 1 = ke^{-m}. \tag{2}$$

Show also, by means of a sketch or otherwise, that equation (2) can have 0, 1 or 2 real roots, depending on the value of k, and find the set of values of k for which such solutions exist. For what set of values of k do the corresponding solutions of (1) tend to zero as $t \to \infty$?

1990 Paper II

Comments

Do not be put off by the words at the very beginning of the question, which you do not need to understand. In any case, the question very soon morphs into an investigation of roots of an equation using curve sketching. However, their meaning can be gleaned from the equation itself (sometimes called a *differential-difference* equation). The left hand side of the equation is the rate of increase of the function f, which may measure the amplitude of some physical disturbance. According to the equation, this is equal to the sum of two terms. One is $-f(t)$, which represents damping: this term alone would give exponentially decreasing solutions. The other is a positive term proportional to $f(t-1)$: this is called a feedback term, because it depends on the value of f one year (say) previously. The feedback term could represent some seasonal effect, while the damping term may be caused by some resistance-to-growth factor.

The suggested method of solving the equation is similar to what you might have used for second order linear equations with constant coefficients: you guess a solution (e^{mt}) and then substitute into the equation to check that this is a possible form of solution and to find the values of m which will work. It is always possible to multiply the exponential by a constant factor, since any constant multiple of a solution is also a solution. (This is a consequence of the *linearity* of the equation: i.e. no terms involving $f(t)^2$, $f(t)^3$, etc.) We can also take linear combinations of solutions to produce a more general solution.

To find the set of values of k which give real solutions of (2), you need to investigate the borderline case, where the two curves $y = m+1$ and $y = ke^{-m}$ just touch (and therefore have the same gradient). Remember that the sign of k is not restricted.

Unlike the second order differential equation case, the equation for m is not quadratic; in fact, it is not even polynomial, since it has an exponential term. This means, as we will find, that there may not be exactly 2 solutions; in general, there may be many solutions (e.g. the non-polynomial equation $\sin m = 0$ has solutions $m = 0, \pi, 2\pi, \ldots$) or no solutions (e.g. $e^m = 0$ has no solutions). And unlike the differential equation case, we have no idea whether the method is going to work, because there may be solutions of a completely different form. The full analysis of equations such as (1) provides an extraordinarily rich field of study with many surprising results.

Solution to problem 44

We do as we are told to start and substitute Ae^{mt} into equation (1). After cancelling the overall factor A, we get

$$me^{mt} = -e^{mt} + ke^{m(t-1)}.$$

We can cancel the overall factor e^{mt}, but an exponential remains on account of the $f(t-1)$ term:

$$m = -1 + ke^{-m},$$

which is equivalent to equation (2) overleaf.

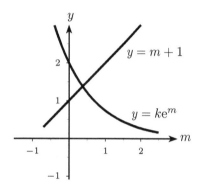
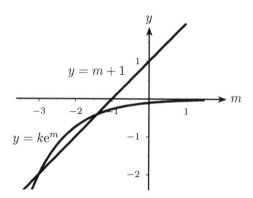

The above sketches show $y = m+1$ and $y = ke^{-m}$ on the same axes, for positive and negative k: in the first sketch, $k = 2$; in the second $k = -0.1$.

As can be seen from the sketches, equation (2) always has exactly one solution if $k > 0$. For $k < 0$, there may be two or zero solutions depending on whether the line and curve intersect, or just one solution if they touch. They will touch if there is a value of m such that

$$m + 1 = ke^{-m} \text{ and } 1 = -ke^{-m},$$

the second of these equations being the condition that the gradients (differentiate with respect to m to find the gradient) of the line and curve are the same at the common point. Solving the two equations gives $m = -2$ and $k = -e^{-2}$.

If $0 > k > -e^{-2}$, then the curve and line will intersect, so the set of values of k for which equation (2) has solutions is $k > 0$ and $-e^{-2} \leqslant k < 0$.

The corresponding solutions to equation (1) tend to zero as $t \to \infty$ if and only if $m < 0$, because then they are exponentially decreasing rather than increasing.

If $k < 0$, we can see from the sketch that the solutions of (2), if there are any (i.e. if $-e^{-2} \leqslant k < 0$), occur in the bottom left quadrant, and so $m < 0$. The corresponding solutions of (1) will tend to zero.

If $k > 0$, the intersection of the graph of $y = ke^{-m}$ with the y-axis is at $y = k$ whereas the intersection of the graph of $y = m + 1$ with the y-axis is at $y = 1$. For $k \geqslant 1$, the solution of (2) occurs in the top right quadrant and has $m \geqslant 0$. For $0 < k < 1$, the solution occurs in the top left quadrant and has $m < 0$.

The range of k for which the solutions tend to zero as $t \to \infty$ is therefore $-e^{-2} \leqslant k < 0$ and $0 < k < 1$.

Problem 45: Lagrange's identity (✓✓✓)

If $y = \mathrm{f}(x)$, the inverse of f is given by *Lagrange's identity*:

$$\mathrm{f}^{-1}(y) = y + \sum_1^\infty \frac{1}{n!} \frac{\mathrm{d}^{n-1}}{\mathrm{d}y^{n-1}} [y - \mathrm{f}(y)]^n,$$

when this series converges.
(i) Verify Lagrange's identity when $\mathrm{f}(x) = ax$.
(ii) Show that one root of the equation $x - \frac{1}{4}x^3 = \frac{3}{4}$ is

$$x = \sum_0^\infty \frac{3^{2n+1}(3n)!}{n!(2n+1)!\, 4^{3n+1}}.\qquad (\dagger)$$

(iii) Find a solution for x, as a series in λ, of the equation $x = \mathrm{e}^{\lambda x}$.
[You may assume that the series in part (ii) converges and that the series in parts (i) and (iii) converge for suitable a and λ.]

1987 Paper II

Comments

This looks pretty frightening at first, because of the complicated and unfamiliar formula. However, its bark is worse than its bite. Once you have decided what you need to find the inverse of, you just substitute it into the formula and see what happens. Do not worry about the use of the word 'convergence'; this can be ignored. It is just included to satisfy the legal eagles who will point out that the series might not have a finite sum.

In part (ii) you can, as it happens, solve the cubic by normal means (find one root by inspection, factorise and use the usual formula to solve the resulting quadratic equation). The root found by Lagrange's equation is the one closest to zero. Equation ($\dagger$) turns out to be a very obscure way of writing a familiar quantity.[29]

Lagrange was one of the leading mathematicians of the 18th century; Napoleon referred to him as the 'lofty pyramid of the mathematical sciences'. He attacked a wide range of problems, from celestial mechanics to number theory. In the course of his investigation of the roots of polynomial equations, he discovered group theory (in particular, his eponymous theorem about the order of a subgroup dividing the order of the group), though the term 'group' and the systematic theory had to wait until Galois and Abel in the first part of the 19th century.

Lagrange's formula, produced before the advent of the theory of integration in the complex plane, which allows a relatively straightforward derivation, testifies to his remarkable mathematical ability. It is practically forgotten now, but in its day it had a great impact. The applications given above give an idea how important it was, in the age before computers.

---

[29] The expansion sums to 1; I don't know how you can see that directly. I thought it would come from using the classical formula for the root of a cubic:

$$-\left[\tfrac{3}{2}\left(1 + i\sqrt{\tfrac{13}{243}}\,\right)\right]^{\tfrac{1}{3}} - \left[\tfrac{3}{2}\left(1 - i\sqrt{\tfrac{13}{243}}\,\right)\right]^{\tfrac{1}{3}},$$

expanding each bracket binomially but it doesn't seem to. The roots are obtained from this complicated expression by noticing that $-\left[\tfrac{3}{2}\left(1 \pm i\sqrt{\tfrac{13}{243}}\,\right)\right]^{\tfrac{1}{3}} = \tfrac{1}{2}\left(1 \pm i\sqrt{\tfrac{13}{3}}\,\right)$.

Solution to problem 45

(i) The inverse of $f(x) = ax$ is given by $f^{-1}(y) = y/a$. Substituting into $f(y) = ay$ into Lagrange's identity gives

$$f^{-1}(y) = y + \sum_1^\infty \frac{1}{n!} \frac{d^{n-1}}{dy^{n-1}} [y - ay]^n = y + \sum_1^\infty (1-a)^n \frac{1}{n!} \frac{d^{n-1} y^n}{dy^{n-1}}$$

$$= y + \sum_1^\infty (1-a)^n y = y + y \frac{1-a}{1-(1-a)},$$

where the last equality follows from summing the geometric progression. This simplifies to y/a, thus verifying Lagrange's formula.

(ii) Let $f(x) = x - \frac{1}{4}x^3$. Then the equation becomes $f(x) = \frac{3}{4}$, so we must find $f^{-1}(\frac{3}{4})$. Again, we just substitute into Lagrange's formula, leaving y arbitrary for the moment:

$$f^{-1}(y) = y + \sum_1^\infty \frac{1}{n!} \frac{d^{n-1}}{dy^{n-1}} [y - (y - \tfrac{1}{4}y^3)]^n = y + \sum_1^\infty \frac{1}{n!} \frac{d^{n-1}}{dy^{n-1}} [\tfrac{1}{4}y^3]^n$$

$$= y + \sum_1^\infty \frac{1}{4^n n!} \frac{d^{n-1}}{dy^{n-1}} y^{3n}$$

$$= y + \sum_1^\infty \frac{1}{4^n n!} \frac{(3n)!}{(2n+1)!} y^{2n+1}.$$

This is a solution to the equation[30] $x - \frac{1}{4}x^3 = y$, so we just set $y = \frac{3}{4}$ to obtain the given result.

(iii) The obvious choice for f is $f(x) = x - e^{\lambda x}$, in which case the equation becomes $f(x) = 0$ and we want $f^{-1}(0)$. Again substituting into Lagrange's identity gives

$$f^{-1}(y) = y + \sum_1^\infty \frac{1}{n!} \frac{d^{n-1}}{dy^{n-1}} \left[e^{\lambda y}\right]^n = y + \sum_1^\infty \frac{1}{n!} \frac{d^{n-1}}{dy^{n-1}} e^{n y \lambda} = y + \sum_1^\infty \frac{1}{n!} (n\lambda)^{n-1} e^{n y \lambda}.$$

Setting $y = 0$ gives a series for the root:

$$f^{-1}(0) = \sum_1^\infty \frac{n^{n-1}}{n!} \lambda^{n-1},$$

which cannot be further simplified.

Post-mortem

Regarding convergence in Lagrange's formula:

Part (i): we need, somewhat mysteriously, $0 < a < 2$ for the geometric progression to converge, but then the result is valid for any y.

Part (ii): Using the approximation $n! \approx (n/e)^n$, which is a simplified version of Stirling's formula, it can be seen that the series converges provided $|y| < 4/\sqrt{27}$. This is related to the condition for x to lie between the two turning points of $f(x)$, which guarantees that $f^{-1}(y)$ is well defined.

Part (iii): We can use Stirling's formula, as above, to show that series converges for $|\lambda| < e^{-1}$. You might like to sketch the two functions x and $e^{\lambda x}$; you should find that the range of values of λ for which the equation has a real root corresponds exactly to the range for which the series converges.

---

[30] Note that this equation cannot in general be solved by spotting roots. In fact, by translation and scaling, any cubic equation can be reduced to this form, so our series solution can be used to find a solution of any cubic ewquation.

Problem 46: Bernoulli polynomials (✓✓)

The Bernoulli polynomials, $B_n(x)$ (where $n = 0, 1, 2, \ldots$), are defined by $B_0(x) = 1$ and, for $n \geq 1$,

$$\frac{dB_n}{dx} = nB_{n-1}(x) \tag{1}$$

and

$$\int_0^1 B_n(x)\,dx = 0. \tag{2}$$

(i) Show that $B_4(x) = x^2(x-1)^2 + c$, where c is a constant (which you need not evaluate).

(ii) Show that, for $n \geq 2$, $B_n(1) - B_n(0) = 0$.

(iii) Show, by induction or otherwise, that

$$B_n(x+1) - B_n(x) = nx^{n-1} \quad (n \geq 1). \tag{3}$$

(iv) Hence show that

$$n \sum_{m=0}^{k} m^{n-1} = B_n(k+1) - B_n(0),$$

and deduce that $\sum_{m=0}^{1000} m^3 = (500500)^2$.

1987 Paper III

Comments

The Swiss family Bernoulli included no fewer than eight mathematicians who were counted amongst the leading scholars of their day. They made major contributions to all branches of mathematics, especially differential calculus. There was great rivalry between some members of the family; between brothers Jakob (1654–1705) and Johann (1667–1748), in particular.

Johann once published an important result in the form of a Latin anagram, in order to retain the priority of discovery without giving the game away to his brother. The anagram was: $24a$, $6b$, $6c$, $8d$, $33e$, $5f$, $2g$, $4h$, $33i$, $6l$, $21m$, $26n$, $16o$, $8p$, $5q$, $17r$, $16s$, $25t$, $32u$, $4x$, $3y$, $+$, $-$, $-$, $\pm$, $=$, 4, 2, 1, '. The notation means that his important result contained, for instance, the letter a 24 times either in text or in equations. After waiting for a year for someone to solve it, Bernoulli weakened and published the solution himself. If you are trying to solve the anagram yourself, you might like to know that it is about the Riccati equation $y' = ay^2 + bx^n$, which can be solved (very cunningly, as it turns out) when n is of the form $-4m/(2m \pm 1)$ for any positive integer m. (Newton also published some work in the form of anagrams, during his conflict with Leibniz).

The polynomials described above were discovered by Jakob Bernoulli. They are defined recursively; that is to say, the zeroth polynomial is given an explicit value, and the nth is determined from the $(n-1)$th. Here, B_{n-1} has to be integrated to obtain B_n, which means that B_n is a polynomial of degree n. The constant of integration is determined by the condition (2), so B_n is uniquely determined. We have to do this explicitly for part (i).

Solution to problem 46

(i) First we find $B_1(x)$ by integrating $1 \times B_0(x)$, using equation (1): $B_1(x) = x + k$, where k is a constant. We find k by applying the condition $\int_0^1 B_1(x)\,dx = 0$, which gives $k = -\frac{1}{2}$. Next we find $B_2(x)$ by similar means, giving $x^2 - x + \frac{1}{6}$, and similarly $B_3(x) = x^3 - \frac{3}{2}x^2 + \frac{1}{2}x$ and $B_4(x)$ is as given.

(ii) We are asked to prove a result involving $B_n(x)$ evaluated at $x = 1$ and $x = 0$, i.e. at the limits of the integral (2). We therefore try the effect of integrating both sides of equation (1) between these limits:

$$B_n(1) - B_n(0) \equiv \int_0^1 \frac{dB_n(x)}{dx}\,dx = n \int_0^1 B_{n-1}(x)\,dx = 0,$$

using property (2) with n replaced by $(n-1)$.

(iii) First the easy bit of the induction proof. For $n = 1$, we have $B_1(x) = x - \frac{1}{2}$, so

$$B_1(x+1) - B_1(x) = (x + 1 - \tfrac{1}{2}) - (x - \tfrac{1}{2}) = 1 \equiv nx^{n-1},$$

so the formula holds.

Now suppose that it holds for $n = k$:

$$B_k(x+1) - B_k(x) - kx^{k-1} = 0 \tag{4}$$

and investigate

$$B_{k+1}(x+1) - B_{k+1}(x) - (k+1)x^k, \tag{5}$$

which we hope will also equal zero.

The only helpful thing we know about Bernoulli polynomials involves the derivatives. Therefore, let us see what happens when we differentiate the expression (5):

$$\frac{d}{dx}B_{k+1}(x+1) - \frac{d}{dx}B_{k+1}(x) - (k+1)kx^{k-1}.$$

Now using (1) gives

$$(k+1)B_k(x+1) - (k+1)B_k(x) - (k+1)kx^{k-1}.$$

Note that we have used the chain rule to differentiate $B_{k+1}(x+1)$ with respect to x rather than with respect to $(x+1)$. Note also that there is a pleasing overall factor of $(k+1)$, which suggests that we are on the right track. In fact, taking out this factor gives exactly the left hand side of equation (4), which is zero.

Of course, we are not finished yet: we have only shown that the derivative of equation (5) is equal to zero; the expression (5) is therefore constant:

$$B_{k+1}(x+1) - B_{k+1}(x) - (k+1)x^k = A.$$

We must show that $A = 0$. Setting $x = 0$ gives $B_{k+1}(1) - B_{k+1}(0) = A$, which implies that $A = 0$ by part (ii).

(iv) Summing (3) from $x = 0$ to $x = k$ gives the first of these results immediately because nearly all the terms cancel in pairs. The evaluation of the sum follows by calculating $B_4(1001) - B_4(0)$ from the result given in part (i).

Problem 47: Vector geometry (✓✓)

The line ℓ has vector equation $\mathbf{r} = \lambda \mathbf{s}$, where

$$\mathbf{s} = (\cos\theta + \sqrt{3})\,\mathbf{i} + \sqrt{2}\,\sin\theta\,\mathbf{j} + (\cos\theta - \sqrt{3})\,\mathbf{k}$$

and λ is a scalar parameter. Find an expression for the angle between ℓ and the line $\mathbf{r} = \mu(a\mathbf{i} + b\mathbf{j} + c\mathbf{k})$. Show that there is a line m through the origin such that, whatever the value of θ, the acute angle between ℓ and m is $\frac{1}{6}\pi$.

A plane has equation $x - z = 4\sqrt{3}$. The line ℓ meets this plane at P. Show that, as θ varies, P describes a circle, with its centre on m. Find the radius of this circle.

2000 Paper II

Comments

It is not easy to set vector questions at this level: they tend to become merely complicated and tedious, rather than difficult in an interesting way. In a good question, there is usually some underlying geometry and it pays to try to understand what this is. Here, the question is about the geometrical object traced out by ℓ as θ varies.

You will need to know about scalar products of vectors for this question, but otherwise it is really just coordinate geometry.

Vectors form an extremely important part of almost every branch of mathematics (maybe *every* branch of mathematics) and will probably be one of the first topics you tackle on your university course.

Solution to problem 47

Both ℓ and the line $\mathbf{r} = \mu(a\mathbf{i}+b\mathbf{j}+c\mathbf{k})$ pass through the origin, so the angle α between the lines is given by the scalar product of the unit vectors, i.e. the scalar product between the given vectors divided by the product of the lengths of the two vectors:

$$\cos\alpha = \frac{a(\cos\theta + \sqrt{3}) + b(\sqrt{2}\sin\theta) + c(\cos\theta - \sqrt{3})}{\sqrt{(\cos\theta + \sqrt{3})^2 + 2\sin^2\theta + (\cos\theta - \sqrt{3})^2}\ \sqrt{a^2 + b^2 + c^2}}$$

$$= \frac{(a+c)\cos\theta + \sqrt{2}\,b\sin\theta + (a-c)\sqrt{3}}{2\sqrt{2}\ \sqrt{a^2+b^2+c^2}}\ . \tag{*}$$

Now we want to show that there is some choice of a, b and c such that $\cos\alpha$ does not depend on the value of θ. By inspection of equation (*), we see that this requires $a = -c$ and $b = 0$.

Setting $a = -c$ in (*) gives $\cos\alpha = \frac{1}{2}\sqrt{3}$ and $\alpha = \frac{1}{6}\pi$ as required. We can absorb the constant a into μ, so the equation of the line m becomes

$$\mathbf{r} = \mu(\mathbf{i} - \mathbf{k})\ . \tag{**}$$

The coordinates of a general point on the line ℓ are

$$x = \lambda(\cos\theta + \sqrt{3}),\quad y = \lambda\sqrt{2}\sin\theta,\quad z = \lambda(\cos\theta - \sqrt{3})\ .$$

For a point which is also on the plane $x - z = 4\sqrt{3}$ we have

$$\lambda(\cos\theta + \sqrt{3}) - \lambda(\cos\theta - \sqrt{3}) = 4\sqrt{3}$$

so $\lambda = 2$. The point P at the intersection between the line and the plane therefore has coordinates

$$(2\cos\theta + 2\sqrt{3},\ 2\sqrt{2}\sin\theta,\ 2\cos\theta - 2\sqrt{3})\ . \tag{***}$$

As θ varies, does P moves round a circle? That is not easy to see, but fortunately we gather from the question that the centre of the circle is on m. It must also lie in the plane of the circle, which is the plane $x - z = 4\sqrt{3}$. The line (**) meets this plane at the point $(2\sqrt{3},\ 0,\ -2\sqrt{3})$; call it O. To verify that P describes a circle with centre O we must check that the distance from P, given by (***), to O is independent of θ. We have

$$OP^2 = (2\cos\theta)^2 + (2\sqrt{2}\sin\theta)^2 + (2\cos\theta)^2 = 8\ ,$$

which is indeed independent of θ. The radius of the circle is therefore $2\sqrt{2}$.

Post-mortem

As mentioned in the comments section, it is helpful to understand the geometry of vector questions. Since the angle between the variable line ℓ and the fixed line m is constant ($\frac{1}{6}\pi$), the shape generated by ℓ as θ varies is the surface of a cone (actually a pair of cones).

The intersection of a plane with a cone is in general a *conic section*: an ellipse (of which a circle is a special case), hyperbola, parabola or pair of straight lines. Try to picture these possibilities. In this case, the normal to the plane, which is in direction $(1, 0, -1)$, is parallel to the axis of the cone (the line m), so the intersection is indeed a circle.

Problem 48: Solving a quartic (✓✓)

Given that
$$x^4 + px^2 + qx + r \equiv (x^2 - ax + b)(x^2 + ax + c),$$

express p, q and r in terms of a, b and c.

Show that a^2 is a root of the cubic equation
$$u^3 + 2pu^2 + (p^2 - 4r)u - q^2 = 0.$$

Verify that $u = 9$ is a root in the case $p = -1$, $q = -6$, $r = 15$.

Hence, or otherwise, solve the equation
$$y^4 - 8y^3 + 23y^2 - 34y + 39 = 0.$$

2000 Paper III

Comments

The long-sought solution of the general cubic was found, in 1535, by Niccolò Tartaglia (c. 1500–1557). He was persuaded to divulge his secret (in the form of a poem) by Girolamo Cardano (1501–1576), who promised not to publish it before Cardano did. However, Cardano discovered that it had previously been discovered by del Ferro (1465–1525/6) before 1515 so he published it himself in his algebra book *The Great Art*. There followed an acrimonious dispute between Tartaglia and Cardano, in which the latter was championed by his student Ferrari (1522–1565). The dispute culminated in a public mathematical duel between Ferrari and Tartaglia held in the church of Santa Maria in Milan in 1548, in which they attempted to solve each others' cubics. The duel ended in a shouting match with Tartaglia storming off. It seems Ferrari was the winner. Tartaglia was sacked from his job as lecturer and Ferrari made his fortune as a tax assessor before becoming a professor of mathematics at Bologna. He was poisoned by his sister, with arsenic, in 1565.

Ferrari found a way of reducing quartic equations to cubic equations; his method (roughly) is used in this question to solve a quartic which could probably be solved easier 'otherwise'. But it is the method that is interesting, not the solution.

The first step of the Ferrari method is to reduce the general quartic to a quartic equation with the cubic term missing by means of a linear transformation of the form $x \to x - a$. Then this reduced quartic is factorised (the first displayed equation in this question). The factorisation can be found by solving a cubic equation (the second displayed equation above) that must be satisfied by one of the coefficients in the factorised form.

The solutions of the quartic are all complex, but don't worry if you haven't come across complex numbers: you will be able to do everything except perhaps write down the last line.

You will no doubt be full of admiration for this clever method of solving quartic equations. One thing you are bound to ask yourself is how the other two roots of the cubic equation fit into the picture. In fact, the cubic equation gives (in general) 6 distinct values of a so there is quite a lot of explaining to do, given that the quartic has at most 4 distinct roots.

Solution to problem 48

We have

$$(x^2 - ax + b)(x^2 + ax + c) = (x^2 - ax)(x^2 + ax) + b(x^2 + ax) + c(x^2 - ax) + bc$$
$$= x^4 + (b + c - a^2)x^2 + a(b - c)x + bc$$

so

$$p = b + c - a^2, \quad q = a(b - c), \quad r = bc. \qquad (*)$$

To obtain an equation for a in terms of p, q and r we eliminate b and c from $(*)$ using the identity $(b + c)^2 = (b - c)^2 + 4bc$. This gives $(p + a^2)^2 = (q/a)^2 + 4r$ which simplifies easily to the given cubic with u replaced by a^2.

We can easily verify by direct substitution that $u = 9$ satisfies the given cubic.

To solve the quartic equation, the first task is to reduce it to the form $x^4 + px^2 + qx + r = 0$, which has no term in x^3. This is done by means of a translation. Noting that $(y - a)^4 = y^4 - 4ay^3 + \cdots$, we set $x = y - 2$. This gives

$$(x + 2)^4 - 8(x + 2)^3 + 23(x + 2)^2 - 34(x + 2) + 39 = 0 .$$

which (not surprisingly) boils down to $x^4 - x^2 - 6x + 15 = 0$, so that $p = -1$, $q = -6$ and $r = 15$. We have already shown that one root of the cubic corresponding to these values is $u = 9$. Thus we can achieve the factorisation of the quartic into two quadratic factors by setting $a = 3$ (or $a = -3$; it doesn't matter which we use) and

$$b = \frac{1}{2}\left(p + a^2 + \frac{q}{a}\right) = 3$$

$$c = \frac{1}{2}\left(p + a^2 - \frac{q}{a}\right) = 5 .$$

Thus

$$x^4 - x^2 - 6x + 15 = (x^2 - 3x + 3)(x^2 + 3x + 5) .$$

Setting each of the two quadratic factors equal to zero gives

$$x = \frac{3 \pm i\sqrt{3}}{2} \quad \text{and} \quad x = \frac{-3 \pm i\sqrt{11}}{2} ,$$

so

$$y = \frac{7 \pm i\sqrt{3}}{2} \quad \text{or} \quad y = \frac{1 \pm i\sqrt{11}}{2} .$$

Post-mortem

Did you work out the relation between the six possible values of a (corresponding to given values of p, q and r) and the roots of the quartic? The point is that the quartic can be written as the product of four linear factors (by the fundamental theorem of algebra) and there are six ways of grouping the linear factors into two quadratic factors. Each way corresponds to a value of a.

Problem 49: Areas and volumes (✓✓)

The function f satisfies the condition $f'(x) > 0$ for $a \leqslant x \leqslant b$, and g is the inverse of f.

(i) By making a suitable change of variable, prove that

$$\int_a^b f(x)\,dx = b\beta - a\alpha - \int_\alpha^\beta g(y)\,dy, \qquad (1)$$

where $\alpha = f(a)$ and $\beta = f(b)$. Interpret this formula geometrically, by means of a sketch, in the case where α and a are both positive.

Verify the result (1) for $f(x) = e^{2x}$, $a = 0$, $b = 1$.

(ii) Prove similarly and interpret the formula

$$2\pi \int_a^b x f(x)\,dx = \pi(b^2\beta - a^2\alpha) - \pi \int_\alpha^\beta [g(y)]^2\,dy. \qquad (2)$$

1987 Paper II

Comments

As is often the case, the required change of variable for part (i) can be worked out by inspection of the limits.

To find the inverse function (note: inverse, not reciprocal) of the function f, it is often best to try to think of the function g such that $g(f(x)) = x$, though making x the subject of $y = f(x)$ is perhaps safer with an unfamiliar function.

The condition $f'(x) > 0$ ensures that f has a unique inverse; a function such as sin which has maximum and minimum points has a unique inverse only on restricted ranges of its argument which do not contain the turning points. (This is obvious from a sketch). The condition $f'(x) < 0$ would do equally well.

The geometrical interpretations of these formulae are exceptionally pleasing, though the second one needs some artistic skill to make it convincing.

Solution to problem 49

(i) The limits of the integral on the right hand side of equation (1) are $f(a)$ and $f(b)$, which suggests the change of variable $y = f(x)$. Making this change, so that $dy = f'(x)dx$, gives

$$\int_\alpha^\beta g(y)\,dy = \int_a^b g(f(x))\,f'(x)\,dx = \int_a^b x\,f'(x)\,dx\,.$$

For the last equality, we have used the definition of g as the inverse of f, i.e. $g(f(x)) = x$. This last integral is begging to be integrated by parts:

$$\int_a^b x\,f'(x)\,dx = x f(x) \Big|_a^b - \int_a^b f(x)\,dx\,,$$

which gives the required result after evaluating $xf(x)$ at a and b. The first sketch below shows these areas: the area between the large and small rectangles is $(b\beta - a\alpha)$, which is split into the areas represented by the two integrals of equation (1), hatched vertically and horizontally, respectively.

Setting $f(x) = e^{2x}$ and $a = 0$, $b = 1$ in the left hand side of (1) gives $\int_0^1 e^{2x}\,dx = \tfrac{1}{2}(e^2 - 1)$. For the right hand side of (1), we have $\alpha = 1$ and $\beta = e^2$ so $b\beta - a\alpha = e^2$. The inverse of e^{2x} is $\tfrac{1}{2}\ln y$ so the integral becomes

$$\int_1^{e^2} \tfrac{1}{2}\ln y\,dy = \tfrac{1}{2}(y\ln y - y)\Big|_1^{e^2} = \tfrac{1}{2}(2e^2 - e^2) - \tfrac{1}{2}(0 - 1) = \tfrac{1}{2}e^2 + \tfrac{1}{2}\,.$$

Thus the left hand side of (1) agrees with the right hand side.

(ii) We can use the same method (change of variable followed by integration by parts):

$$\int_\alpha^\beta [g(y)]^2\,dy = \int_a^b x^2 f'(x)\,dx = x^2 f(x)\Big|_a^b - \int_a^b 2x f(x)\,dx = (b^2\beta - a^2\alpha) - 2\int_a^b x f(x)\,dx\,,$$

which gives the required formula (2) on multiplication by π.

The first of the integrals in (2) is the volume of the solid body *under* the surface formed by rotating the curve $y = f(x)$ round the y-axis; this volume is thought of as a set of concentric cylindrical shells of height $f(x)$ with internal radius x ($a \leqslant x \leqslant b$), and thickness dx. The second integral is the volume *inside* the surface formed by rotating the curve $y = f(x)$ round the y-axis; this volume is thought of as a pile of infinitesimally thin discs of radius $g(y)$ ($\alpha \leqslant y \leqslant \beta$) and thickness dy. The sum of the two integrals is equal to the difference between the volumes of the two concentric cylinders (of radii a and b, heights α and β, respectively) as shown in the second sketch below.

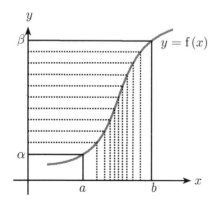
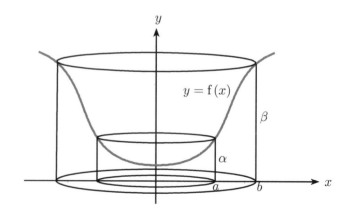

Problem 50: More curve sketching (✓✓✓)

(i) The curve C_1 passes through the origin in the x–y plane and its gradient is given by

$$\frac{dy}{dx} = x(1-x^2)e^{-x^2}.$$

Show that C_1 has a minimum point at the origin and a maximum point at $\left(1, \tfrac{1}{2}e^{-1}\right)$. Find the coordinates of the other stationary point. Give a rough sketch of C_1.

(ii) The curve C_2 passes through the origin and its gradient is given by

$$\frac{dy}{dx} = x(1-x^2)e^{-x^3}.$$

Show that C_2 has a minimum point at the origin and a maximum point at $(1, k)$, where $k > \tfrac{1}{2}e^{-1}$. (You need not find k.)

2001 Paper II

Comments

No work is required to find the x coordinate of the stationary points, but you have to integrate the differential equation to find the y coordinate. For the second part, you cannot integrate the equation — other than numerically, or in terms of rather obscure special functions that you almost certainly haven't come across, such as the incomplete gamma function defined by

$$\Gamma(x, a) = \int_0^a t^{x-1} e^{-t} dt.$$

However, you can obtain an estimate, which is all that is required, by comparing the gradients of C_1 with C_2 and thinking of the graphs for $-1 \leqslant x \leqslant 1$. This is perhaps a bit tricky; an idea that you may well not alight on under examination conditions.

Solution to problem 50

C_1 has stationary points when $\dfrac{dy}{dx} = 0$, i.e. when $x = 0$, $x = +1$ or $x = -1$. To find the y coordinates of the stationary points, we integrate the differential equation, using integration by parts:

$$y = \int (1-x^2)xe^{-x^2}\,dx = (1-x^2)(-\tfrac{1}{2}e^{-x^2}) - \int xe^{-x^2}\,dx = -\tfrac{1}{2}(1-x^2)e^{-x^2} + \tfrac{1}{2}e^{-x^2} + \text{const} = \tfrac{1}{2}x^2 e^{-x^2},$$

where we have used the fact that C_1 passes through the origin to evaluate the constant of integration. The coordinates of the stationary points are therefore $(1, \tfrac{1}{2}e^{-1})$, $(0,0)$ and $(-1, \tfrac{1}{2}e^{-1})$.

One way of classifying the stationary points is to look at the second derivative:

$$\dfrac{dy}{dx} = (x - x^3)e^{-x^2} \implies \dfrac{d^2 y}{dx^2} = (1 - 3x^2)e^{-x^2} - 2x^2(1 - x^2)e^{-x^2} = (1 - 5x^2 + 2x^4)e^{-x^2},$$

which is positive when $x = 0$ (indicating a minimum) and negative when $x = 1$ and $x = -1$ (indicating maxima).

Since $y \to 0$ as $x \to \pm\infty$, the sketch is:

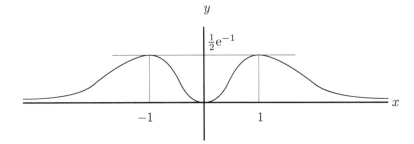

For C_2, the stationary points are again at $x = 0$ and $x = \pm 1$. To classify the stationary points, we calculate the second derivative:

$$\dfrac{d^2 y}{dx^2} = (1 - 3x^2 - 3x^3 + 3x^5)e^{-x^3},$$

which is positive at $x = 0$ (a minimum) and negative at $x = 1$ (a maximum).

Since we cannot integrate this differential equation explicitly, we must compare C_2 with C_1 to see whether the value of y at the maxima is indeed greater for C_2 than for C_1.

For $0 < x < 1$, $x^3 < x^2$, so $e^{x^3} < e^{x^2}$ and $e^{-x^3} > e^{-x^2}$. Thus the gradient of C_2 is greater than the gradient of C_1. Since both curves pass through the origin, we deduce that C_2 lies above C_1 for $0 < x \leqslant 1$ and therefore the maximum point $(1, k)$ on C_2 has $k > \tfrac{1}{2}e^{-1}$.

Post-mortem

You were probably surprised that the examiners didn't ask you to sketch C_2; they usually do. It looks as if it could be done, because you know that C_2 passes through the origin and you can relate its slope, roughly at least, to that of C_1, which you did sketch. The difficulty lies in determining its behaviour as $x \to \infty$; does $y \to 0$ as for C_1? In fact, it doesn't; no reason why it should. Instead, $y \to 0.153...$. We shouldn't have expected C_1 to asymptote to the x-axis either; the fact that it does so is due to a rather delicate balancing of the two terms that determine its gradient.

Problem 51: Spherical loaf (✓✓✓)

A spherical loaf of bread is cut into parallel slices of equal thickness. Show that, after any number of the slices have been eaten, the area of crust remaining is proportional to the number of slices remaining.

A European ruling decrees that a parallel-sliced spherical loaf can only be referred to as 'crusty' if the ratio of volume V (in cubic metres) of bread remaining to area A (in square metres) of crust remaining after any number of slices have been eaten satisfies $V < A$. Show that the radius of a crusty parallel-sliced spherical loaf must be less than $2\frac{2}{3}$ metres.

[The area A and volume V formed by rotating a curve in the x–y plane round the x-axis from $x = a - t$ to $x = a$ are given by

$$A = 2\pi \int_{a-t}^{a} y\left(1 + \left(\frac{dy}{dx}\right)^2\right)^{\frac{1}{2}} dx, \qquad V = \pi \int_{a-t}^{a} y^2 \, dx \, .]$$

2001 Paper I

Comments

The first result (the mathematical result, I mean, not the European ruling which I invented) came as a bit of a surprise to me — though no doubt it is well known. I wondered if it was the only surface of revolution with this property. You might like to think about this.

Don't be distracted by the use of the word 'slices'; since the thickness of the slices is not given, it is clear that you are supposed to think in terms of the continuous distance along the loaf rather than the number of slices.

For the last part, you will need to minimise a ratio as a function of t (the 'length' of loaf remaining). To find the ratio you have to do a couple of integrals. It is this 'multi-stepping' that makes the problem difficult (and very different from typical school-level questions) rather than any individual step.

Solution to problem 51

The first thing we need is an equation for the surface of a spherical loaf. The obvious choice, especially given the hint at the bottom of the question, is the circle $x^2 + y^2 = a^2$ in the plane $z = 0$, rotated about the x-axis.

If the loaf is cut at a distance t from the end $x = a$, and the portion from $x = -a$ to $x = a - t$ is eaten, then the area remaining is

$$2\pi \int_{a-t}^{a} y\left(1 + \left(\frac{dy}{dx}\right)^2\right)^{\frac{1}{2}} dx = 2\pi \int_{a-t}^{a} (a^2 - x^2)^{\frac{1}{2}} \left(1 + \left(\frac{-x}{(a^2 - x^2)^{\frac{1}{2}}}\right)^2\right)^{\frac{1}{2}} dx$$

$$= 2\pi \int_{a-t}^{a} a\, dx$$

$$= 2\pi a t. \tag{$*$}$$

This is proportional to the length t of remaining loaf, so proportional to the number of slices remaining (if the loaf is evenly sliced).

The remaining volume is

$$\pi \int_{a-t}^{a} y^2 dx = \pi \int_{a-t}^{a} (a^2 - x^2)\, dx = \pi \left[(a^2 x - \tfrac{1}{3}x^3)\right]_{a-t}^{a}$$

$$= \pi(a t^2 - \tfrac{1}{3}t^3).$$

As a quick check on the algebra, notice that this is zero when $t = 0$ and $\tfrac{4}{3}\pi a^3$ when $t = 2a$.

Thus $V/A = (3at - t^2)/(6a)$. This is a quadratic curve with zeros at $t = 0$ and $t = 3a$, so it has a maximum at $t = \tfrac{3}{2}a$ (by differentiating or otherwise), where $V/A = \tfrac{3}{8}a$. Since we require this ratio to be less than one, we must have $a < \tfrac{8}{3}$ metres.

Post-mortem

It was well worth studying the information given at the end of the question before plunging into the question: it not only gave the necessary formulae for the surface area and volume, but also gave them in a form that suggested a way forward right at the start of the question.

Did you think about whether there are other shapes that would have the property proved for the sphere in the first paragraph of the question? Mathematically, it boils down to whether there are functions $y(x)$, other than our function $y = \sqrt{a^2 - x^2}$, that can satisfy $(*)$. If we differentiate $(*)$, we obtain

$$2\pi y \left(1 + \left(\frac{dy}{dx}\right)^2\right)^{\frac{1}{2}} = 2\pi a t$$

which looks formidable, but in fact simplifies to an equation that you can integrate quite easily. The sphere is, it turns out, the only shape with the required property.

Problem 52: Snowploughing (✓✓✓)

> Two identical snowploughs plough the same stretch of road. The first starts at a time t_1 seconds after it starts snowing, and the second starts from the same point $t_2 - t_1$ seconds later, going in the same direction. Snow falls so that the depth of snow increases at a constant rate of k ms^{-1}. The speed of each snowplough is ak/z ms^{-1} where z is the depth (in metres) of the snow it is ploughing and a is a constant. Each snowplough clears all the snow. Show that the time t at which the second snowplough has travelled a distance x metres satisfies the equation
>
> $$a\frac{\mathrm{d}t}{\mathrm{d}x} = t - t_1 e^{x/a}. \quad (\dagger)$$
>
> Hence show that the snowploughs will collide when they have travelled $a(t_2/t_1 - 1)$ metres.

1987 Specimen Paper III

Comments

There is something exceptionally beautiful about this question, but it is hard to identify exactly what it is; seeing the question for the first time makes even hardened mathematicians smile with pleasure.

There is a modelling element to it: you have to set up equations from the information given in the text. The first equation you need is a simple first order differential equation to find the time taken by the first snowplough to travel a distance x. The corresponding equation for the second snowplough is a bit more complicated, because the depth of snow at any point depends on the time at which the first snowplough reached that point, clearing the snow.

The differential equation ($\dagger$) can be solved using an integrating factor. However, the equation which arises naturally at this point is one involving $\dfrac{\mathrm{d}x}{\mathrm{d}t}$, which cannot (apparently) be solved by any means. It is the rather good trick of turning the equation upside down (regarding t as a function of x instead of x as a function of t) that allows the problem to be solved so neatly. Apologies if you haven't come across integrating factors for first order differential equations; they are not on our syllabus, but they are really not difficult — you can look online and find an easily understandable explanation.

You won't surprised to learn that there is a generalisation to n identical snowploughs . . . .

Solution to problem 52

Suppose that the first snowplough reaches a distance x at time T after it starts snowing. Then the depth of snow it encounters is kT and its speed is therefore $ak/(kT)$, i.e. a/T. The equation of motion of the first snowplough is

$$\frac{dx}{dT} = \frac{a}{T}.$$

Integrating both sides with respect to T gives

$$x = a \ln T + \text{constant of integration.}$$

We know that $T = t_1$ when $x = 0$ (the snowplough started t_1 seconds after the snow started), so

$$x = a \ln T - a \ln t_1.$$

This can be rewritten as

$$T = t_1 e^{x/a}.$$

When the second snowplough reaches x at time t, snow has been falling for a time $t - T$ since it was cleared by the first snowplough, so the depth at time t is $k(t - T)$ metres, i.e. $k(t - t_1 e^{x/a})$ metres. Thus the equation of motion of the second snowplough is

$$\frac{dx}{dt} = \frac{ak}{k(t - t_1 e^{x/a})}.$$

Now we use the standard result (a special case of the chain rule)

$$\frac{dx}{dt} = 1 \bigg/ \frac{dt}{dx}$$

to obtain the required equation (†).

Multiplying by $e^{-x/a}$ (an *integrating factor*) and rearranging gives

$$e^{-x/a} \frac{dt}{dx} - \frac{e^{-x/a} t}{a} = -\frac{t_1}{a} \quad \text{i.e.} \quad \frac{d}{dx}\left(t e^{-x/a}\right) = -\frac{t_1}{a}$$

which integrates to

$$t e^{-x/a} = -\frac{t_1}{a} x + \text{constant of integration.}$$

Since the second snowplough started ($x = 0$) at time t_2, the constant of integration is just t_2 and the solution is

$$t = (t_2 - t_1 x/a) e^{x/a}.$$

The snowploughs collide when they reach the same position at the same time. Let this position be $x = X$. Then

$$T = t \implies t_1 e^{X/a} = (t_2 - t_1 X/a) e^{X/a},$$

so X is given by

$$t_1 = (t_2 - t_1 X/a).$$

This is equivalent to the given formula.

Problem 53: Tortoise and hare (✓✓)

A tortoise and a hare have a race to the vegetable patch, a distance X kilometres from the starting post, and back. The tortoise sets off immediately, at a steady v kilometres per hour. The hare goes to sleep for half an hour and then sets off at a steady speed V kilometres per hour. The hare overtakes the tortoise half a kilometre from the starting post, and continues on to the vegetable patch, where she has another half an hour's sleep before setting off for the return journey at her previous pace. One and quarter kilometres from the vegetable patch, she passes the tortoise, still plodding gallantly and steadily towards the vegetable patch. Show that

$$V = \frac{10}{4X - 9}$$

and find v in terms of X.

Find X if the hare arrives back at the starting post one and a half hours after the start of the race.

How long does it take the tortoise to reach the vegetable patch?

Comments

The first thing to do is draw a distance–time diagram. You might also find it useful to let the times at which the two animals meet be T_1 and T_2.

Solution to problem 53

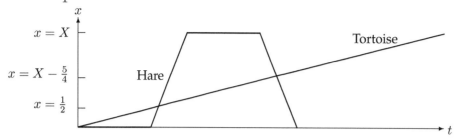

Let the times of the first and second meetings be T_1 and T_2. Then

$$vT_1 = \tfrac{1}{2}, \quad V(T_1 - \tfrac{1}{2}) = \tfrac{1}{2}, \quad vT_2 = (X - \tfrac{5}{4}), \quad V(T_2 - 1) = X + \tfrac{5}{4}.$$

The first pair and the second pair of equations give, respectively,

$$\frac{1}{v} - \frac{1}{V} = 1, \qquad \frac{X - \tfrac{5}{4}}{v} = \frac{X + \tfrac{5}{4}}{V} + 1,$$

and hence (first eliminating v):

$$V = \frac{10}{4X - 9}, \qquad v = \frac{10}{4X + 1}.$$

The total distance travelled by the hare in $\tfrac{1}{2}$ hour is $2X$, so

$$2X = \frac{10}{4X - 9} \times \frac{1}{2}, \quad \text{i.e.} \quad 8X^2 - 18X - 5 = 0.$$

Factorising the quadratic gives two roots $-\tfrac{1}{4}$ and $\tfrac{5}{2}$, and we obviously need the positive root. The speed of the tortoise is given by

$$v = \frac{10}{4X + 1} = \frac{10}{11}$$

so the time taken to travel $\tfrac{5}{2}$ kilometres is $\tfrac{11}{4}$ hours.

Post-mortem

It is interesting to see how quickly a mechanics question can change from mechanics to something else: algebra, calculus or, as in this case, coordinate geometry. As soon as the picture is drawn, the equations of the lines can be written down and everything follows as if it were a geometry problem.

Sometimes the change happens even more quickly. There is a well-known problem about a monk ascending a hill one day and descending the next day starting at the same time on each day. The problem is to show that there is a time at which the monk was at the same height on both days. You are not even given the speed at which the monk climbs. But if you draw (or just imagine drawing) height-time graphs of the ascent and descent, and superimpose them, you are done.

Problem 54: How did the chicken cross the road? (✓✓)

A single stream of cars, each of width a and exactly in line, is passing along a straight road of breadth b with speed V. The distance between successive cars (i.e. the distance between back of one car and the front of the following car) is c.

A chicken crosses the road in safety at a constant speed u in a straight line making an angle θ with the direction of traffic. Show that

$$u \geqslant \frac{Va}{c\sin\theta + a\cos\theta}. \qquad (*)$$

Show also that if the chicken chooses θ and u so that she crosses the road at the least possible (constant) speed, she crosses in time

$$\frac{b}{V}\left(\frac{c}{a} + \frac{a}{c}\right).$$

1997 Paper I

Comments

I like this question because it relates to (an idealised version of) a situation we have probably all thought about. Once you have visualised it, there are no great difficulties. As usual, you have to be careful with the inequalities, though it turns out here that there is no danger of dividing by a negative quantity.

Solution to problem 54

The easiest way to think about this problem is to consider the cars to be stationary and the velocity of the chicken to be $(u\cos\theta - V, u\sin\theta)$. Then the diagrams are very easy to visualise.

Let t be the time taken to cross the distance a in which the chicken is at risk. Then $a = ut\sin\theta$.

For safety, the chicken must choose $ut\cos\theta + c \geqslant Vt$: equality here occurs when the chicken starts at the near-side rear of one car and just avoids being hit by the far-side front of the next car.

Eliminating t from these two equations gives the required inequality:

$$ut\cos\theta \geqslant Vt - c$$
$$\Longrightarrow (u\cos\theta - V)t \geqslant -c$$
$$\Longrightarrow (u\cos\theta - V)\frac{a}{u\sin\theta} \geqslant -c$$
$$\Longrightarrow au\cos\theta - aV \geqslant -cu\sin\theta$$
$$\Longrightarrow u(c\sin\theta + a\cos\theta) \geqslant aV$$

which is the required result.

For a given value of θ, the minimum speed satisfies

$$u(c\sin\theta + a\cos\theta) = aV.$$

The smallest value of this u is therefore obtained when $c\sin\theta + a\cos\theta$ is largest. This can be found by calculus (regard it as a function of θ and differentiate: the maximum occurs when $\tan\theta = c/a$) or by trigonometry:

$$c\sin\theta + a\cos\theta = \sqrt{a^2 + c^2}\cos(\theta - \arctan(c/a))$$

so the maximum value is $\sqrt{a^2 + c^2}$ and it occurs when $\tan\theta = c/a$.

The time of crossing is

$$\frac{b}{u\sin\theta} = \frac{b(c\sin\theta + a\cos\theta)}{Va\sin\theta} = \frac{b(c + a\cot\theta)}{Va} = \frac{b(c + a^2/c)}{Va}.$$

Post-mortem

There is another inequality besides (∗) that you might have noticed. If $u\cos\theta > V$ (so the chicken moves faster than the cars — a bit unlikely unless the chicken is trying to cross the M25), the chicken should start her run at the *front* nearside of a car and must not collide with the car *ahead*. This requires $(u\cos\theta - V)t \leqslant c$, so

$$u(-c\sin\theta + a\cos\theta) \leqslant aV.$$

If $(-c\sin\theta + a\cos\theta) < 0$, this places no constraint on u. But if $(-c\sin\theta + a\cos\theta) > 0$, then

$$u \leqslant \frac{aV}{-c\sin\theta + a\cos\theta}.$$

In both cases, the inequality (∗) overleaf does not apply. This is clearly not the situation envisaged by the examiners, and probably not by any of the candidates either, but still it should have been catered for in the wording of the question.

Problem 55: Hank's gold mine (✓✓)

Hank's Gold Mine has a very long vertical shaft of height l. A light chain of length l passes over a small smooth light fixed pulley at the top of the shaft. To one end of the chain is attached a bucket A of negligible mass and to the other a bucket B of mass m.

The system is used to raise ore from the mine as follows. When bucket A is at the top it is filled with mass $2m$ of water and bucket B is filled with mass λm of ore, where $0 < \lambda < 1$. The buckets are then released, so that bucket A descends and bucket B ascends. When bucket B reaches the top both buckets are emptied and released, so that bucket B descends and bucket A ascends. The time to fill and empty the buckets is negligible. Find the time taken from the moment bucket A is released at the top until the first time it reaches the top again.

This process goes on for a very long time. Show that, if the greatest amount of ore is to be raised in that time, then λ must satisfy the condition $f'(\lambda) = 0$ where

$$f(\lambda) = \frac{\lambda(1-\lambda)^{1/2}}{(1-\lambda)^{1/2} + (3+\lambda)^{1/2}}.$$

1998 Paper I

Comments

One way of working out the acceleration of a system of two masses connected by a light string passing over a pulley is to write down the equation of motion of each mass, bearing in mind that the force due to tension will be the same for each mass (it cannot vary along the string, because then the acceleration of some portion of the *massless* string would be infinite). Then you eliminate the tension.

Alternatively, you can use the equation of motion of the system of two joined masses. The system has inertial mass equal to the sum of the masses (because both masses must accelerate equally) but gravitational mass equal to the difference of the masses (because the gravitational force on one mass cancels, partially, the gravitational force on the other), so the equation of motion is just (Newton's law of motion)

$$(m_1 + m_2)a = (m_1 - m_2)g.$$

Solution to problem 55

When bucket A ascends, the acceleration is g.

For bucket A's downward journey, at acceleration a, the equations of motion for bucket A and bucket B, respectively, are
$$-T + 2mg = 2ma, \quad T - (1+\lambda)mg = (1+\lambda)ma,$$
where T is the tension in the rope. Eliminating T gives so $a = \dfrac{1-\lambda}{3+\lambda} g$.

The time of descent (using $l = \tfrac{1}{2}at^2$) is $\sqrt{2l/a}$ and the time of ascent is $\sqrt{2l/g}$. The total time required for one complete cycle is therefore
$$\sqrt{\frac{2l}{g}}\left(1 + \sqrt{\frac{3+\lambda}{1-\lambda}}\right).$$

Call this t. The number of round trips in a long time t_{long} is t_{long}/t so the amount of ore lifted in time t_{long} is $\lambda m t_{\text{long}}/t$.

To maximise this, we have to maximise λ/t with respect to λ, and λ/t is exactly the $\text{f}(\lambda)$ given. Note that $\text{f}(0) = 0$ (which makes sense because no ore is raised if $\lambda = 0$), and $\text{f}(1) = 0$ (which also makes sense because the buckets don't move on the raising stage if $\lambda = 1$). That means the greatest value of $\text{f}(\lambda)$ must occur at a value of λ in the range $0 < \lambda < 1$ at which $\text{f}'(\lambda) = 0$.

Post-mortem

You may have wondered why, in the last part, the question says that the process goes on for a very long time. The reason for this is that when you maximise what I have called $\lambda m t_{\text{long}}/t$ the result may not correspond to a complete number of cycles. If you stop in mid-cycle, you raise no ore from that cycle so a calculus maximisation of a continuous function is not the right method. However, if the process continues for many cycles, the contribution from the last cycle becomes negligible, and a calculus maximisation becomes appropriate.

You may also wonder why you were not asked to find the maximising value of λ. If you are feeling exceptionally energetic, you could try to solve $\text{f}'(\lambda) = 0$. This will eventually lead you to the rather discouraging quartic equation
$$\lambda^4 + 4\lambda^3 + 2\lambda^2 - 8\lambda + 3 = 0.$$

You will have done a lot of tedious work to obtain an equation that cannot be solved without either a formula for the solutions of a general quartic equation or a computer. There are in fact two real solutions of this equation, $\lambda = 0.528$ and $\lambda = 0.704$.

Problem 56: A chocolate orange (✓)

A chocolate orange consists of a sphere of delicious smooth uniform chocolate of mass M and radius a, sliced into segments by planes through a fixed axis. It stands on a horizontal table with this axis vertical and it is held together by a narrow ribbon round its equator. Show that the tension in the ribbon is at least $\frac{3}{32}Mg$.

[You may assume that the centre of mass of a segment of angle 2θ is at distance $\dfrac{3\pi a \sin\theta}{16\theta}$ from the axis.]

1987 Specimen Paper II

Comments

This question can be done by the usual methods (resolving forces and taking moments about a suitably chosen point). Since the chocolate is smooth, there is no friction. The ribbon may be elastic, so it could be tighter than is needed just to keep the orange together. At the lowest tension possible the orange is on the point of falling apart, so there are no forces between the faces of the segments, except at the point of contact with the table.

The ribbon is said to be thin, but actually this has to be interpreted as massless as well as thin; a ribbon with mass would simply drop off the equator of the orange.

Notice that the question does not say that the segments are all the same size. That suggests that we should look at just one segment, expecting to find that the tension required for that segment is independent of the angle of the segment.

When I set this question originally on a 1985 examination paper (pre-dating STEP, which started in 1987), I gave the wrong formula for the distance of the centre of mass of the segment, and the answer for the tension was also incorrect, but consistent with the incorrect formula. Not surprisingly, no one pointed it out, either at the time or afterwards. With a bit of luck, it is correct now.

I sent a copy of the examination paper to a well-known manufacturer of high quality chocolate confection[31] and was rewarded with a puzzled letter and a parcel; not as large a parcel as I had hoped for (perhaps they spotted the incorrect formula), but better than nothing. It seemed worth including the question in this book in case they wanted an opportunity to make amends ....

---

[31] Now located in Poland. Originally, the firm made chocolate apples which, being apple-shaped, would not have worked well for this question (the centre of mass of an orange segment was hard enough for me). The chocolate lemon, introduced in 1974, would not have been good either but luckily it was discontinued rapidly after it turned out that no one wanted to eat it.

Solution to problem 56

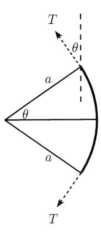

The above diagram shows a horizontal cross-section of a segment of the orange.

Resolving the tension T in the ribbon as shown gives a horizontal component of force due to tension on the segment of $2T\sin\theta$ towards the left of the diagram.

The volume of the segment is a fraction $2\theta/(2\pi)$ of the volume of the sphere, so the segment has mass $M(2\theta)/(2\pi)$. The weight of the segment gives a force of $Mg\theta/\pi$ acting downwards through the centre of mass as shown in the diagram below, which is a vertical cross-section of a segment.

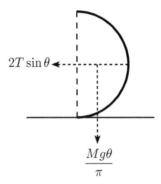

Taking moments about the point of contact of the table and the segment gives

$$\frac{Mg\theta}{\pi} \times \frac{3\pi a \sin\theta}{16\theta} = 2T\sin\theta \times a$$

which gives the required answer.

Post-mortem

Another way of tackling this sort of problem is to use the principle of virtual work, for which you imagine that the system relaxes a very small amount (in this case, by allowing the ribbon to stretch) and equate the work done against the constraints (here, tension times extension) to the change of potential energy of the system, to obtain a formula for the constraint force (here, the tension). In many cases, this method is simpler, but here it turns out to be very difficult: not recommended at all. The difficulty is that if the orange relaxes so that each segment is now inclined at a small angle α to the vertical, quite a lot of work is needed to calculate how far the centre of mass has fallen, even working to lowest order in α.

Problem 57: Lorry on bend

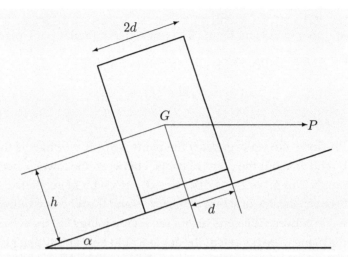

A lorry of weight W stands on a plane inclined at an angle α to the horizontal. Its wheels are a distance $2d$ apart, and its centre of gravity G is at a distance h from the plane, and halfway between the sides of the lorry. A horizontal force P acts on the lorry through G, as shown.

(i) If the normal reactions on the lower and higher wheels of the lorry are equal, show that the sum of the frictional forces between the wheels and the ground is zero.

(ii) If P is such that the lorry does not tip over (but the normal reactions on the lower and higher wheels of the lorry need not be equal), show that

$$W \tan(\alpha - \beta) \leqslant P \leqslant W \tan(\alpha + \beta),$$

where $\tan \beta = d/h$.

2002 Paper I

Comments

There is not much more to this than just resolving forces and taking moments about a suitable point.

You might think it a bit odd to have a force that acts horizontally through the centre of gravity of the lorry: it is supposed to be centrifugal. The first draft of the question was intended as a model of a lorry going round a bend on a cambered road. The idea was to relate the speed of the lorry to the angle of camber: the speed should be chosen so that there is be no tendency to skid. However, the modelling part of the question was eventually abandoned.

You can do the second part algebraically (using equations derived from resolving forces and taking moments), but there is a more direct approach.

There was a slight inaccuracy in the wording of the question, which I have not corrected here. Maybe you will spot it in the course of your solution. The diagram is correct.

Solution to problem 57

Let the normal reactions at the lower and upper wheels be N_1 and N_2, respectively, and let the frictional forces at the lower and upper wheels be F_1 and F_2, respectively, (both up the plane).

(i) Taking moments about G, we have

$$(N_1 - N_2)d = (F_1 + F_2)h \qquad (*)$$

so if $N_1 = N_2$ the sum of the frictional forces is zero.

(ii) The condition for the lorry not tip over *down* the plane is $N_2 \geqslant 0$, which is the same as saying that the total moment of W and P about the point of contact between the lower wheel and the plane in the clockwise sense is positive. This gives $P \leqslant W \tan(\alpha + \beta)$ after a bit of geometry. Actually, working out the shortest distance between the line of action of the forces and the wheels requires a bit of thought. You find that $\alpha + \beta$ is the angle between the vertical and the lines joining G to the wheels. This is obvious, at least in retrospect: if the plane were not tilted, the angle would be β and tilting just increases this by the angle of tilt.

The other inequality comes from a similar calculation using the other wheel.

Post-mortem

I expect that you found that really clear and BIG diagrams made the question much easier: that is nearly always the case with mechanics.

The slight inaccuracy in the question occurs because if the line of action of P passes between the wheels of the lorry (not, as shown in the diagram, higher than the higher wheel), then the lorry cannot tip over up the plane however large P is. The given inequality is not correct in this case, which corresponds to $\alpha + \beta > \frac{1}{2}\pi$ and therefore $\tan(\alpha + \beta) < 0$.

An alternative way of obtaining the inequalities of part (ii) is to work with the resultant of P and W. The direction of this resultant force must be such that its line of action passes between the wheels. The line of action makes an angle of $\tan^{-1}(P/W)$ with the vertical through G and the inequalities follow after the same geometry as in the answer to part (ii) above.

Problem 58: Fielding (✓✓)

> In a game of cricket, a fielder is perfectly placed to catch a ball. She watches the ball in flight and takes the catch just in front of her eye. The angle between the horizontal and her line of sight at a time t after the ball is struck is θ. Show that $\dfrac{d}{dt}(\tan\theta)$ is constant during the flight.
>
> The next ball is also struck in the direction of the fielder but at a different velocity. In order to be perfectly placed to catch the ball, the fielder runs at constant speed towards the batsman. Assuming that the ground is horizontal, show that again $\dfrac{d}{dt}(\tan\theta)$ is constant during the flight.

✓✓

Comments

As with all the very best questions, nine-tenths of this question is submerged below the surface. It uses the deepest properties of Newtonian dynamics, and a good understanding of the subject makes the question completely transparent. However, it can still be done without too much trouble by a straightforward approach, in which case the difficulty lies only in setting it up for yourself.

The cleverness of this question lies in its use of two fundamental invariances of Newton's second law:

$$m\frac{d^2\mathbf{x}}{dt^2} = m\mathbf{g}. \tag{†}$$

The first is invariance under *time reflection symmetry*, which arises because equation (†) is not affected by the transformation $t \to -t$. This means that any given solution can be replaced by one where the projectile goes back along the trajectory, i.e. time runs backwards.

The second is invariance under what are called *Galilean transformations*. Equation (†) is also invariant under the transformation $\mathbf{x} \to \mathbf{x} + \mathbf{v}t$, where $\mathbf{v}$ is an arbitrary constant velocity. This means that we can solve the equation in a frame that moves with constant speed.

Solution to problem 58

We take a straightforward approach. Let the height above the fielder's eye-level at which the ball is struck be h. Let the speed at which the ball is struck be u and the angle which the trajectory of the ball initially makes with the horizontal be α.

Then, taking x to be the horizontal distance of the ball at time t from the point at which the ball was struck and y to be the height of the ball at time t above the fielder's eye-level[32], we have

$$x = (u\cos\alpha)t\,,\quad y = h + (u\sin\alpha)t - \tfrac{1}{2}gt^2\,.$$

Let d be the horizontal distance of the fielder from the point at which the ball is struck, and let T be the time of flight of the ball. Then

$$d = (u\cos\alpha)T\,,\quad 0 = h + (u\sin\alpha)T - \tfrac{1}{2}gT^2 \qquad (*)$$

and

$$\tan\theta = \frac{y}{d-x} = \frac{h + (u\sin\alpha)t - \tfrac{1}{2}gt^2}{d - (u\cos\alpha)t}$$

$$= \frac{-(u\sin\alpha)T + \tfrac{1}{2}gT^2 + (u\sin\alpha)t - \tfrac{1}{2}gt^2}{(u\cos\alpha)T - (u\cos\alpha)t} \qquad \text{(using *)}$$

$$= \frac{-(u\sin\alpha)(T-t) + \tfrac{1}{2}g(T^2 - t^2)}{(u\cos\alpha)(T-t)}$$

$$= \frac{-u\sin\alpha + \tfrac{1}{2}g(T+t)}{u\cos\alpha}\,. \qquad \text{(cancelling the factor } (T-t)\text{)}$$

This last expression is a polynomial of degree one in t, so its derivative is constant, as required.

For the second part, let l be the distance from the fielder's original position to the point at which she catches the ball. Then $l = vT$ and

$$\tan\theta = \frac{y}{(l-vt) + d - x} = \frac{y}{v(T-t) + d - x} = \frac{-u\sin\alpha + \tfrac{1}{2}g(T+t)}{v + u\cos\alpha}$$

cancelling the factor of $(T-t)$ as before. This again has constant derivative.

Post-mortem

The invariance mentioned in the comments section above can be used to answer the question almost without calculation.

Using time reflection symmetry to reverse the trajectory shows that the batsman is completely irrelevant: it only matters that the fielder caught a ball. We just think of the ball being projected from the fielder's hands (the time-reverse of a catch). Taking her hands as the origin of coordinates, and using u to denote the projection (i.e. the catching) speed of the ball and α to be the angle of projection (i.e. the final value of θ), we have $y = (u\sin\alpha)t - \tfrac{1}{2}gt^2$, $x = (u\cos\alpha)t$ and $\tan\theta = (u\sin\alpha - \tfrac{1}{2}gt)/u\cos\alpha$. The first derivative of this expression is constant, as before.

We use the Galilean transformation for the second part of the question. Instead of thinking of the fielder running with constant speed v towards the batsman, we can think of the fielder being stationary and the ball having an additional horizontal speed of v. The situation is therefore not changed from that of the first part of the question, except that $u\cos\theta$ should be replaced by $v + u\cos\theta$.

---

[32] Draw a diagram! I would, but there isn't enough room on the page.

Problem 59: Equilibrium of rod of non-uniform density (✓✓)

A rigid straight rod AB has length l and weight W. Its weight per unit length at a distance x from B is

$$\alpha W l^{-1} \left(\frac{x}{l}\right)^{\alpha-1},$$

where α is a constant greater than 1. Show that the centre of mass of the rod is at a distance $\dfrac{\alpha l}{\alpha+1}$ from B.

The rod is placed with the end A on a rough horizontal floor and the end B resting against a rough vertical wall. The rod is in a vertical plane at right angles to the plane of the wall and makes an angle of θ with the floor. The coefficient of friction between the floor and the rod is μ and the coefficient of friction between the wall and the rod is also μ. Show that, if the equilibrium is limiting at both A and B, then

$$\tan\theta = \frac{1-\alpha\mu^2}{(1+\alpha)\mu}.$$

Given that $\alpha = \frac{3}{2}$ and given also that the rod slides for any $\theta < \frac{1}{4}\pi$ find the greatest possible value of μ.

2002 Paper II

Comments

This is a pretty standard situation: a rod leaning against a wall, prevented from slipping by friction at both ends. The only slight variation is that the rod is not uniform; the only effect of this is to alter the position of the centre of gravity through which the weight of the rod acts.

The question is not general, since the coefficient of friction is the same at both ends of the rod. I think that it is a good idea (at least, if you are not working under examination conditions) to set out the general equations (with coefficients of friction μ_A and μ_B and frictional forces F_A and F_B) without at first assuming limiting friction. This will help understand the structure of the equations and highlight the symmetry between the ends of the rod.

You might wonder why the weight per unit length is given in such a complicated way; why not simply kx^α? The reason is to keep the dimensions honest. The factor of W/l appears so that the dimension is clearly weight divided by length. Then x appears only divided by l to make a dimensionless ratio. Finally, the factor of α is required for the total weight of the rod to be W.

Solution to problem 59

The distance, $\bar{x}$, of the centre of mass from the end B of the rod is given by

$$\bar{x} = \frac{\int_0^l \alpha W l^{-\alpha} x^\alpha \, dx}{\int_0^l \alpha W l^{-\alpha} x^{\alpha-1} \, dx} = \frac{\int_0^l x^\alpha \, dx}{\int_0^l x^{\alpha-1} \, dx} = \frac{(\alpha+1)^{-1} l^{\alpha+1}}{(\alpha)^{-1} l^\alpha} = \frac{\alpha l}{\alpha+1}.$$

The response to the rest of this question should be preceded by an annotated diagram showing all relevant forces on the rod and with G nearer to A than B. This will help to clarify ideas.

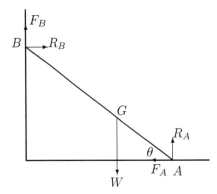

The equations that determine the equilibrium are

$$F_B + R_A = W \qquad \text{(vertical forces)}$$
$$F_A - R_B = 0 \qquad \text{(horizontal forces)}$$
$$R_B \bar{x} \sin\theta + F_B \bar{x} \cos\theta = R_A(l-\bar{x})\cos\theta - F_A(l-\bar{x})\sin\theta . \qquad \text{(clockwise moments about } G\text{)}$$

If the rod is about to slip then the frictional and normal forces at A can be specified as $F_A = \mu R_A$ and $F_B = \mu R_B$. Substituting in first two of the above equations gives

$$R_A = \frac{W}{\mu^2+1}, \quad R_B = \frac{\mu W}{\mu^2+1}, \quad F_A = \frac{\mu W}{\mu^2+1}, \quad F_B = \frac{\mu^2 W}{\mu^2+1}.$$

Substituting into the third equation gives the required result.

For the final part, setting $\alpha = \frac{3}{2}$ gives

$$\tan\theta = \frac{2-3\mu^2}{5\mu}$$

If the rod slips for any angle less than $\theta = \frac{1}{4}\pi$, then the angle at which limiting friction occurs at both ends must be at least $\frac{1}{4}\pi$. Therefore

$$\frac{2-3\mu^2}{5\mu} \geqslant 1$$

i.e. $3\mu^2 + 5\mu - 2 \leqslant 0$, or $(3\mu-1)(\mu+2) \leqslant 0$. The greatest possible value of μ is $\frac{1}{3}$.

Post-mortem

In the original version, the question spoke about limiting friction without saying that it was limiting at both ends. We then wondered whether the friction could be limiting at one end only. The answer is yes. The three equations that govern the system do not determine uniquely the four unknown forces (normal reaction and frictional forces at each end of the rod); extra information is required. If the angle is also to be determined, two extra pieces of information are required; in this case the information is that the friction is limiting at both ends.

Problem 60: Newton's cradle (✓✓)

N particles P_1, P_2, P_3, ..., P_N with masses m, qm, q^2m, ..., $q^{N-1}m$, respectively, are at rest at distinct points along a straight line in gravity-free space. The particle P_1 is set in motion towards P_2 with velocity V and in every subsequent impact the coefficient of restitution is e, where $0 < e < 1$.

(i) Show that after the first impact the velocities of P_1 and P_2 are

$$\left(\frac{1-eq}{1+q}\right)V \quad \text{and} \quad \left(\frac{1+e}{1+q}\right)V,$$

respectively.

(ii) Show that, if $q \leqslant e$, then there are exactly $N-1$ impacts.

(iii) Show further that, if $q = e$, then the total loss of kinetic energy after all impacts have occurred is

$$\tfrac{1}{2}me(1 - e^{N-1})V^2.$$

1999 Paper II

Comments

This situation models the toy called 'Newton's Cradle' which consists of four or more heavy metal balls suspended from a frame so that they can swing. At rest, they are in contact in a line. When the first ball is raised and let swing, there follows a rather pleasing pattern of impacts. In this case, the coefficient of restitution is nearly 1 and the balls all have the same mass, so, as the first displayed formula shows, the impacting ball is reduced to rest by the impact. At the first swing of the ball, nothing happens except that the first ball is reduced to rest and the last ball swings away. Note that this is consistent with the balls being separated by a very small amount; what actually happens is that the ball undergoes a small elastic deformation at the impact, and the impulse takes a small amount of time to be transmitted across the ball to the next ball.

Solution to problem 60

(i) Let V_1 and V_2 be the velocities of P_1 and P_2 after the first collision. Using conservation of momentum and Newton's law of impact at the first collision results in the equations

$$mV = mV_1 + mqV_2, \quad V_1 - V_2 = -eV,$$

so

$$V_1 = \left(\frac{1-eq}{1+q}\right)V \quad \text{and} \quad V_2 = \left(\frac{1+e}{1+q}\right)V.$$

(ii) Note that V_1 is positive since $eq \leqslant e^2 < 1$. Conserving momentum and using Newton's law (and doing a bit of algebra) shows that the speeds $\widehat{V}_2$ and V_3 of P_2 and P_3 after the next collision are given by

$$\widehat{V}_2 = \left(\frac{1-eq}{1+q}\right)V_2 = \left(\frac{1-eq}{1+q}\right)\left(\frac{1+e}{1+q}\right)V \quad \text{and} \quad V_3 = \left(\frac{1+e}{1+q}\right)V_2 = \left(\frac{1+e}{1+q}\right)^2 V. \quad (*)$$

Note that $\widehat{V}_2 \geqslant V_1$ so there is no further collision between P_1 and P_2. Applying this argument at each collision shows that there are exactly $N-1$ collisions: P_1 with P_2; P_2 with P_3; etc.

(iii) The speed of P_k after it has hit P_{k+1} is (by extending $(*)$)

$$\left(\frac{1-eq}{1+q}\right)\left(\frac{1+e}{1+q}\right)^{k-1} V,$$

and the speed of P_{k+1} after this collision and before it hits P_{k+2} is

$$\left(\frac{1+e}{1+q}\right)^k V.$$

If $q = e$,

$$V_1 = \left(\frac{1-e^2}{1+e}\right)V = (1-e)V$$

and similarly the final speeds of $P_2, \ldots, P_{N-1}$ are all $(1-e)V$. The final speed of P_N is V. Thus the final total kinetic energy is

$$\tfrac{1}{2}(m + mq + \cdots + mq^{N-2})[(1-e)V]^2 + \tfrac{1}{2}mq^{N-1}V^2 \;=\; \tfrac{1}{2}m\frac{1-e^{N-1}}{1-e}(1-e)^2 V^2 + \tfrac{1}{2}me^{N-1}V^2$$
$$= \tfrac{1}{2}m(1 - e + e^N)V^2,$$

(replacing all the q's with e's and summing the geometric progression). Thus the loss of kinetic energy is

$$\tfrac{1}{2}mV^2 - \tfrac{1}{2}m(1 - e + e^N)V^2,$$

as required.

Post-mortem

There are two tricky aspects to these multiple collision questions. First there is the matter of notation. Above, I have used a hat for the second collision of a particle ($\widehat{V}_2$), retaining the subscript for labelling particles. That works, but if P_2 undergoes another collision, entailing a double hat, it starts getting messy. You could use a different letter, but that gets confusing. The only good method is to use a double subscript: $V_{m,n}$ is the velocity of the mth particle after the nth collision. But that is a sledge hammer for this nut of a question.

The other tricky aspect is getting the signs right. I always think of velocity (not speed), and it is always positive for particles travelling to the right. I also use common sense to check each equation! As usual, a good diagram of each collision is essential.

Problem 61: Kinematics of rotating target (✓✓)

An automated mobile dummy target for gunnery practice is moving anti-clockwise around the circumference of a large circle of radius R in a horizontal plane at a constant angular speed ω. A shell is fired from O, the centre of this circle, with initial speed V and angle of elevation α.

(i) Show that if $V^2 < gR$, then no matter what the value of α, or what vertical plane the shell is fired in, the shell cannot hit the target.

(ii) Assume now that $V^2 > gR$ and that the shell hits the target, and let β be the (positive) angle through which the target rotates between the time at which the shell is fired and the time of impact. Show that β satisfies the equation

$$g^2\beta^4 - 4\omega^2 V^2 \beta^2 + 4R^2\omega^4 = 0.$$

Deduce that there are exactly two possible values of β.

(iii) Let β_1 and β_2 be the possible values of β and let P_1 and P_2 be the corresponding points of impact. By considering the quantities $(\beta_1^2 + \beta_2^2)$ and $\beta_1^2\beta_2^2$, or otherwise, show that the linear distance between P_1 and P_2 is

$$2R\sin\left(\frac{\omega}{g}\sqrt{V^2 - Rg}\right).$$

1999 Paper II

Comments

The rotation of the target is irrelevant for the first part, which contravenes the setters' rule of not introducing information before it is required. In this case, it seemed better to describe the set-up immediately — especially as you are asked for a familiar result.

Remember, when you are considering the roots of the quartic (which is really a quadratic in β^2), that the question gives $\beta > 0$.

The hint in the last paragraph ('by considering ...') is supposed to direct you towards the relation between the coefficients in a quadratic equation and the sum and product of the roots; otherwise, you get into some pretty heavy algebra.

Solution to problem 61

(i) This part is about the range of the gun. If the shell lands at distance x, then

$$x = (V\cos\alpha)t, \quad 0 = (V\sin\alpha)t - \tfrac{1}{2}gt^2.$$

Eliminating t gives $xg = V^2\sin 2\alpha$. The range r is the largest value of x, which given by $rg = V^2$. The shell cannot reach the target (for any angle of elevation) if $R > r$, so the shell cannot hit its target if $V^2 < Rg$.

(ii) The time of flight is $R/(V\cos\alpha)$, and this must also equal the time for the target to rotate through β, i.e. β/ω. Thus $\cos\alpha = R\omega/(\beta V)$. Substituting this into the range equation $Rg = 2V^2\sin\alpha\cos\alpha$ gives

$$Rg = 2V^2 \frac{R\omega}{\beta V}\left(1 - \frac{R^2\omega^2}{\beta^2 V^2}\right)^{\frac{1}{2}}.$$

Squaring both sides of the equation, and simplifying, leads to the given quadratic in β^2. Solving the quadratic using the quadratic formula gives

$$g^2\beta^2 = 2\omega^2\left(V^2 \pm \sqrt{V^4 - g^2 R^2}\right).$$

Since $V^2 > gR$, both the values of β^2 roots are real and positive, but they lead to only two relevant values of β since we only want positive values.

(iii) The linear distance between the two points of impact is $2R\sin\tfrac{1}{2}(\beta_2 - \beta_1)$. Now $\beta_1^2 + \beta_2^2 = 4\omega^2 V^2/g^2$ and $\beta_1^2\beta_2^2 = 4R^2\omega^4/g^2$ (sum and product of the roots of the quadratic), so

$$(\beta_2 - \beta_1)^2 = \beta_2^2 + \beta_1^2 - 2\beta_1\beta_2 = \frac{4\omega^2 V^2}{g^2} - \frac{4R\omega^2}{g},$$

which leads to the given result.

Post-mortem

Why do we get two values for β? In order to hit the large circle, if it is in range, we can fix any speed of projection and choose the angle of projection appropriately. In general, there are two possible angles of projection, one above $45°$ and one below $45°$ (satisfying $\sin 2\alpha = Rg/V^2$).

However, in this question, we want to hit the circle essentially at a given time (when the target is at the landing place), which means that $V\cos\alpha$ is fixed, though of course it is a little more complicated because the position of the target is fixed but unknown. But still, the two values of β, corresponding to two different angles of elevation, emerge in the same way.

Did you notice the slight inaccuracy in the question? The final 'distance' could be negative (if for example $\pi g < \omega\sqrt{V^2 - Rg} < 2\pi g$). Mod signs are needed.

Problem 62: Particle on wedge (✓✓✓)

A wedge of mass M rests on a smooth horizontal surface. The face of the wedge is a smooth plane inclined at an angle α to the horizontal. A particle of mass m slides down the face of the wedge, starting from rest. At a later time t, the speed V of the wedge, the speed v of the particle and the angle β of the velocity of the particle below the horizontal are as shown in the diagram.

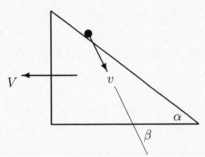

Let y be the vertical distance descended by the particle. Derive the following results, stating in (ii) and (iii) the mechanical principles you use:

(i) $V \sin \alpha = v \sin(\beta - \alpha)$;

(ii) $\tan \beta = (1+k) \tan \alpha$, where $k = m/M$;

(iii) $2gy = v^2(1 + k \cos^2 \beta)$.

Hence show that $2g'y = \left(\dfrac{dy}{dt}\right)^2$, where $g' = \dfrac{g \tan^2 \beta}{\tan^2 \beta + K} = \dfrac{gK \tan^2 \alpha}{K \tan^2 \alpha + 1}$ and $K = 1 + k$. Deduce that $y = \tfrac{1}{2} g' t^2$.

1998 Paper II

Comments

I was surprised at the difficulty of this problem, compared with a particle on a fixed wedge. There is an easier method (just using forces and Newton's second law) but the route suggested here uses basic principles, the outcomes of which are the numbered parts of the question. For a fixed wedge, the horizontal component of momentum is not conserved, because of the force required to hold the wedge; but (i) and (iii) are just what you would use in the fixed case.

Part (ii) is used only (apart from the very last result) to show that the angle β is constant. This is a bit at first sight surprising: it means that the particle moves in a straight line.

Most of the important intermediate results are given to you, but it is still very good discipline to check that they hold in special cases: for example, $k = 0$ corresponding to a massless particle or equivalently a fixed wedge; and $\alpha = 0$ or $\alpha = \tfrac{1}{2}\pi$ corresponding to a horizontal or vertical wedge face. You should check that you understand what should happen in these special cases and that your understanding is consistent with the formulae.

Solution to problem 62

(i) This equation follows immediately from the constraint placed on the particle: it remains in contact with the plane so components of the velocities of the particle and the wedge perpendicular to the face must be equal.

(ii) Horizontal momentum is conserved, so $MV = mv \cos \beta$. Substitution for V/v using (i) gives

$$k \cos \beta = \frac{V}{v} = \frac{\sin(\beta - \alpha)}{\sin \alpha} = \frac{\sin \beta \cos \alpha - \sin \alpha \cos \beta}{\sin \alpha} = \frac{\sin \beta}{\tan \alpha} - \cos \beta$$

which leads immediately to the required result. Note that the angle β remains constant in the motion.

(iii) Conservation of energy gives

$$mgy = \tfrac{1}{2}MV^2 + \tfrac{1}{2}mv^2, \quad \text{i.e.} \quad kgy = \tfrac{1}{2}(kv \cos \beta)^2 + \tfrac{1}{2}kv^2.$$

Thus $2gy = (k \cos^2 \beta + 1)v^2$.

The next step is to write v in terms of y to obtain the differential equation. The vertical component of velocity is $v \sin \beta$, so

$$v \sin \beta = \frac{dy}{dt}.$$

Now using the result of (iii) gives

$$\frac{dy}{dt} = v \sin \beta = \sqrt{\frac{2gy}{1 + k \cos^2 \beta}} \sin \beta = \sqrt{\frac{2gy \sin^2 \beta}{1 + (K-1) \cos^2 \beta}} = (2g')^{\tfrac{1}{2}} y^{\tfrac{1}{2}}. \qquad (*)$$

Squaring gives the required result, and we can use part (ii) to obtain the expression for g' in terms of α and K.

Now, at last, we use the hard-won result that β, and hence g', is constant. The equation

$$\left(\frac{dy}{dt}\right)^2 = 2g'y$$

is exactly the same as the equation of conservation of energy for a particle falling vertically in a gravitational field of strength g'. For such a particle, we have $y = \tfrac{1}{2}g't^2$, as required.

Post-mortem

The basic principles behind this solution are conservation of momentum and energy, together with the constraint that the particle moves on the surface of the face of the wedge. The benefit of using conservation of energy is that you do not have to worry about the normal reaction force, because it does no work.

Of course, we could obtain the final result by integrating the differential equation $(*)$, which takes only a few lines.

Problem 63: Sphere on step (✓)

A uniform solid sphere of radius a and mass m is drawn very slowly and without slipping from horizontal ground onto a step of height $\frac{1}{2}a$ by a horizontal force of magnitude F which is always applied to the highest point of the sphere and is always perpendicular to the vertical plane which forms the face of the step. Find the maximum value of F in the motion, and prove that the coefficient of friction between the sphere and the edge of the step must exceed $1/\sqrt{3}$.

1997 Paper II

Comments

This is quite straightforward once you have realised that *very slowly* means so slowly that the sphere can be considered to be static at each position. You just have to solve a statics problem with the usual tools: resolving forces and taking moments about suitably chosen points.

It seems odd to have a force that always acts at the highest point of the sphere. The idea is that there is a string wrapped round the sphere and it is being pulled horizontally to get the sphere up the step.

Solution to problem 63

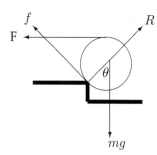

Let the angle between the radius to the point of contact with the step and the downward vertical be θ, as shown. At the point of contact, let the frictional force, which is tangent to the sphere, be f and the reaction R (along the radius of the sphere).

To investigate F, we take moments about a point cunningly chosen to eliminate other unknown forces. Since the lines of action both R and f pass through the point of contact of the sphere with the step, we can eliminate both forces by taking moments about this point:

$$Fa(1 + \cos\theta) = mga\sin\theta,$$

so

$$F = mg\frac{\sin\theta}{1+\cos\theta} = mg\tan\tfrac{1}{2}\theta.$$

This takes its maximum value when θ is largest, i.e. when the sphere just touches the horizontal ground. At this position, $\cos\theta = \tfrac{1}{2}$ and $\sin\theta = \tfrac{1}{2}\sqrt{3}$ and $F_{\max} = mg/\sqrt{3}$.

Taking moments about the centre of the sphere (to eliminate the normal reaction and weight) gives $F = f$, and resolving forces parallel to the radius at the point of contact gives

$$R = F\sin\theta + mg\cos\theta.$$

Now

$$F\sin\theta = mg\frac{\sin^2\theta}{1+\cos\theta} = mg\frac{1-\cos^2\theta}{1+\cos\theta} = mg(1-\cos\theta)$$

so $R = mg$. We therefore need $\mu mg > F_{\max}$, as required.

Post-mortem

Should we have known that $R = mg$ (in the last line of the solution)? It doesn't look obvious from the diagram. However, if we follow the principle, used twice already, of taking moments about a point that eliminates unwanted forces, the result drops out. To relate mg and R directly, we need to eliminate F and f. We therefore need to take moments about the intersection of the lines of action of these two forces, which is shown in the diagram (call it P). Moments must balance about *any* point: it doesn't matter whether the point is inside the body, or outside it as in the case of P. It is easy to see from the geometry (f and F act along tangents to the sphere) that the distances from P to the lines of action of mg and R are equal, which means that the two forces are equal.

To be precise about the meaning of *slow* in this context, you have to compare the dynamic forces connected with the motion of the sphere with the static forces. The motion of the sphere is rotation about the fixed point of contact with the step. If we assume that the centre moves with constant speed, the only extra force due to the motion is an additional reaction at the point of contact with the step. This reaction is centrifugal in nature and so is roughly of the form mv^2/a, which is to be small compared with R. The static approximation ('very slowly') is therefore valid if $v^2 \ll ga$.

Problem 64: Elastic band on cylinder (✓)

A smooth cylinder with circular cross-section of radius a is held with its axis horizontal. A light elastic band of unstretched length $2\pi a$ and modulus of elasticity λ is wrapped round the circumference of the cylinder, so that it forms a circle in a plane perpendicular to the axis of the cylinder. A particle of mass m is then attached to the rubber band at its lowest point and released from rest.

(i) Given that the particle falls to a distance $2a$ below the axis of the cylinder, but no further, show that
$$\lambda = \frac{9\pi mg}{(3\sqrt{3} - \pi)^2}.$$

(ii) Given instead that the particle reaches its maximum speed at a distance $2a$ below the axis of the cylinder, find a similar expression for λ.

2001 Paper I

Comments

This question uses the most basic ideas in mechanics, such as conservation of energy. I included it in this selection of questions without realising that the properties of stretched strings are not in the syllabus (which is given in the appendix). However, I didn't throw it out: it is a nice question and the only two things you need to know about stretched strings are, for a stretched string of natural (i.e. unstretched) length l and extended length $l + x$ with modulus of elasticity λ:

(i) the potential energy stored in the stretched string is $\dfrac{\lambda x^2}{2l}$;

(ii) the tension in the stretched string, by Hooke's law, is $\dfrac{\lambda x}{l}$.

Solution to problem 64

The diagram below shows the system when the particle has fallen a distance a from its initial position.

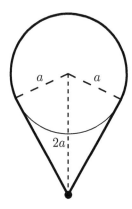

A bit of geometry (including Pythagoras) on the above diagram shows that the length of the extended band is $2\pi a - 2a \arccos \frac{1}{2} + 2\sqrt{3}\,a$ so the extension of the band is

$$-2a \arccos \tfrac{1}{2} + 2\sqrt{3}\,a, \quad \text{i.e.} \quad 2a(\sqrt{3} - \tfrac{1}{3}\pi).$$

(i) At the lowest point, the speed is zero. That suggests using an energy equation, which will involve speed and displacement (for the potential energy) but not acceleration.

Conserving energy (taking the initial potential energy to be zero), we find that when the particle has fallen a distance a and has speed v,

$$0 = \tfrac{1}{2}mv^2 + \tfrac{1}{2}\lambda \frac{[2a(\sqrt{3} - \tfrac{1}{3}\pi)]^2}{2\pi a} - mga.$$

At the lowest point, $v = 0$, which gives the required answer.

(ii) Now we are interested in the point where the speed is greatest, i.e. where the acceleration is zero, so this time we should use the equation of motion of the particle.

The component of the tension T in the band in the vertical direction, acting upwards on the particle, when the particle has fallen a distance y is $2T \cos \theta$ where $\sin \theta = a/(a+y)$. Applying Newton's second law to the motion of the particle gives

$$m\ddot{y} = 2T \cos \theta - mg, \qquad (*)$$

where, by Hooke's law, $T = \lambda \times \text{extension}/2\pi a$. At the maximum speed, $\ddot{y} = 0$. This occurs at $y = a$, so $(*)$ becomes

$$0 = 2\lambda \frac{2a(\sqrt{3} - \tfrac{1}{3}\pi)}{2\pi a} \frac{\sqrt{3}}{2} - mg.$$

In this case $\lambda = \dfrac{\sqrt{3}\,\pi mg}{3\sqrt{3} - \pi}$.

Post-mortem

The question was relatively simple because you only needed to evaluate energy or tension at very special points: maximum extension or maximum speed. You could of course have worked from the general equation of motion $(*)$, with $\cos \theta = \sqrt{2ay + y^2}/(a+y)$ and extension $2\sqrt{2ay + y^2} - 2a \arcsin\left(a/(a+y)\right)$, but you can't solve the differential equation to find the general motion. It would be possible for small oscillations.

Problem 65: A knock-out tournament (✓)

A tennis tournament is arranged for 2^n players. It is organised as a knockout tournament, so that only the winners in any given round proceed to the next round. Opponents in each round except the final are drawn at random, and in any match either player has a probability $\frac{1}{2}$ of winning. Two players are chosen at random before the start of the first round. Find the probabilities that they play each other:

(i) in the first round;

(ii) in the final round;

(iii) in the tournament.

1987 Specimen Paper II

Comments

Note that the set-up is not the usual one for a tennis tournament, where the only random element is in the first round line-up. Two players cannot then meet in the final if they are in the same half of the draw.

Part (i) is straightforward, but parts (ii) and (iii) need a bit of thought. There is a short way and a long way of tackling these parts, and both have merits.

It is a good plan to check your answers, if possible, by reference to simple special cases where you can see what the answers should be; $n = 1$ or $n = 2$, for example.

Interestingly, the answers are independent of the probability that the players have of winning a match; the 2s in the answers represent the number of players in each match rather than (the reciprocal of) the probability that each player has of winning a match. It also does not matter how the draw for each round is made. This is clear if you use the short method mentioned above.

Solution to problem 65

Call the two players P_1 and P_2.

(i) Once P_1 has been given a slot, there are $2^n - 1$ slots for P_2, in only one of which will he or she play P_1. The probability of P_1 playing P_2 is therefore

$$\frac{1}{2^n - 1}.$$

Note that this works for $n = 1$ and $n = 2$.

(ii) **Long way**. To meet in the final, P_1 and P_2 must each win every round before the final, and must also not meet before the final. The probability that P_1 and P_2 do not meet in the first round and that they both win their first round matches, is

$$\left(1 - \frac{1}{2^n - 1}\right)\frac{1}{2^2} \quad \text{i.e.} \quad \frac{1}{2}\left(\frac{2^{n-1} - 1}{2^n - 1}\right).$$

The probability that they win each round and do not meet before the final (i.e. for $n - 1$ rounds) is

$$\frac{1}{2}\left(\frac{2^{n-1} - 1}{2^n - 1}\right) \times \frac{1}{2}\left(\frac{2^{n-2} - 1}{2^{n-1} - 1}\right) \times \cdots \times \frac{1}{2}\left(\frac{2^1 - 1}{2^2 - 1}\right) \quad \text{i.e.} \quad \frac{1}{2^{n-1}}\frac{1}{2^n - 1}.$$

(ii) **Short way**. Since all processes are random here, the probability that any one pair contests the final is the same as that for any other pair. There are a total of $\frac{1}{2} \times 2^n(2^n - 1)$ different pairs, so the probability for any given pair is $1/[2^{n-1}(2^n - 1)]$.

(iii) **Long way**. We need to add the probabilities that P_1 and P_2 meet in each round. The probability that they meet in the kth round is the probability that they reach the kth round times the probability that they meet in the kth round given that they reach it, the latter (conditional) probability being $1/(2^{n-k+1} - 1)$, as can be inferred from part (i). As in part (ii), the probability that they reach the kth round is

$$\frac{1}{2^{k-1}}\frac{2^{n-k+1} - 1}{2^n - 1},$$

so the probability that they meet in the kth round is

$$\frac{1}{2^{k-1}}\frac{2^{n-k+1} - 1}{2^n - 1} \times \frac{1}{2^{n-k+1} - 1} = \frac{1}{2^{k-1}}\frac{1}{2^n - 1}.$$

Summing this as a geometric progression from $k = 1$ to n gives $1/2^{n-1}$.

(iii) **Short way**. By the same short argument as in part (ii), the probability of a given pair meeting in any given match (not necessarily the final) is $1/[2^{n-1}(2^n - 1)]$. Since the total number of matches is $2^n - 1$ (because one match is needed to knock out each player, and all players except one get knocked out), the probability of a given pair playing is

$$\frac{2^n - 1}{2^{n-1}(2^n - 1)} = \frac{1}{2^{n-1}}.$$

Post-mortem

If (like me) you plodded through this question the long way, you might be wondering how you were supposed to think of the short way. Instead of working out what happens to individual players as they progress through the tournament, you think about the space of all possible outcomes (the *sample space*), and attach a probability to each. That way, you can use the symmetry between all the players to help you.

Problem 66: Harry the calculating horse (✓✓)

> Harry the Calculating Horse will do any mathematical problem I set him, providing the answer is 1, 2, 3 or 4. When I set him a problem, he places a hoof on a large grid consisting of unit squares and his answer is the number of squares partly covered by his hoof. Harry has circular hoofs, of radius $\frac{1}{4}$ unit.
>
> After many years of collaboration, I suspect that Harry no longer bothers to do the calculations, instead merely placing his hoof on the grid completely at random. I often ask him to divide 4 by 4, but only about $\frac{1}{4}$ of his answers are right; I often ask him to add 2 and 2, but disappointingly only about $\frac{1}{16}\pi$ of his answers are right. Is this consistent with my suspicions?
>
> I decide to investigate further by setting Harry many problems, the answers to which are 1, 2, 3, or 4 with equal frequency. If Harry is placing his hoof at random, find the expected value of his answers. The average of Harry's answers turns out to be 2. Should I get a new horse?

1997 Paper II

Comments

Hans von Osten, a horse, lived in Berlin around the turn of the last century. He was known far and wide for his ability to solve complex arithmetical problems. Distinguished scientists travelled to Berlin to examine Hans and test his marvellous ability. They would write an equation on a chalkboard and Hans would respond by pawing the ground with his hoof. When Hans reached the answer he would stop. Though he sometimes made errors, his success rate was far higher than would be expected if his answers were random. The accepted verdict was that Hans could do arithmetic.

Hans's reputation as a calculating horse nosedived when an astute scientist simply made sure that neither the person asking the questions nor the audience knew the answers. Hans became an instant failure. His success was based on his ability to sense any change in the audience: a lifted eyebrow, a sigh, a nodding head or the tensing of muscles was enough to stop him from pawing the ground. Anyone who knew the answer was likely to give almost imperceptible clues to the horse. But we shouldn't overlook Hans's talents: at least he had terrific examination technique.

We are investigating the situation when Harry places his hoof at random, so that the probability of the centre of his hoof lands in any given region is proportional to the area of the region. We therefore move swiftly from a question about probability to a question about areas. You just have to divide a given square into regions, each determined by the number of squares that will be partially covered by Harry's hoof if its centre lands in the region under consideration. Remember that the areas must add to one, so the most difficult area calculation can be left until last and deduced from the others.

To answer the last part properly, you really need to set out a hypothesis testing argument: you will accept the null hypothesis (random hoof placing) if, using the distribution implied by the null hypothesis, the probability of obtaining the given result is greater than some pre-decided figure. Obviously, nothing so elaborate was intended here, since it is the last demand of an already long question: just one line would do.

Solution to problem 66

The diagram shows one square of the grid.

Harry's hoof will land completely within exactly one square if he places the centre of his hoof in a square of side $\frac{1}{2}$ centred on the centre of any square, shown with horizontal and vertical lines in the diagram. The area of any such square is $\frac{1}{4}$ square units.

His hoof will partially cover exactly four squares if he places the centre in a circle of radius $\frac{1}{4}$ centred on any intersection of grid lines. The total area of any one such circle is $\frac{1}{16}\pi$ square units, which may be thought of as four quarter circles, one in each corner of any given square, shown lightly shaded in the diagram.

His hoof will partially cover exactly two squares if he places the centre in any one of four $\frac{1}{2}$ by $\frac{1}{4}$ rectangles, of total area $4 \times \frac{1}{8}$ in any given square, shown darkly shaded in the diagram.

Otherwise, his hoof will partially cover three squares; the area of this remaining region, shown with no shading in the diagram, is
$$1 - \frac{1}{4} - \frac{\pi}{16} - \frac{1}{2}.$$

If Harry placed his hoof at random, the probabilities of the different outcomes would be equal to the corresponding area calculated above (divided by the total area of the square, which is 1). Thus the data given in the question are exactly consistent with random placement.

The expected value given random placements is
$$1 \times \frac{1}{4} + 2 \times \frac{1}{2} + 3 \times \left(1 - \frac{1}{4} - \frac{1}{2} - \frac{\pi}{16}\right) + 4 \times \frac{\pi}{16} = 2 + \frac{\pi}{16}$$

The expected value (given that Harry gets all questions right) is $(1+2+3+4)/4 = 5/2$. Harry has a less accurate expected value even than the random expected value. He is clearly hopeless and should go.

Post-mortem

The explanations in the first paragraph of the solution are very concise. To understand them, you should draw yourself lots of diagrams.

The proportion of Harry's answers that are correct, in the case of the $2+2$ calculation is given as 'about $\frac{1}{16}\pi$'. That should have given you pause for thought. Why not $\frac{1}{5}$? Clearly, it was supposed to be helpful, so it should have been no surprise that the probability of getting the answer 4 under the random assumption turned out to be exactly $\frac{1}{16}\pi$. Of course, these questions shouldn't turn into a game of spotting hidden clues, but it is always worth looking for unusual features of a question — or of any mathematical problem.

Problem 67: PIN guessing (✓)

In order to get money from a cash dispenser I have to punch in a Personal Identification Number. I have forgotten my PIN, but I do know that it is equally likely to be any one of the integers 1, 2, ..., n. I plan to punch in integers in ascending order until I get the right one. I can do this at the rate of r integers per minute. As soon as I punch in the first wrong number, the police will be alerted. The probability that they will arrive within a time t minutes is $1 - e^{-\lambda t}$, where λ is a positive constant. If I follow my plan, show that the probability of the police arriving before I get my money is

$$\sum_{k=1}^{n} \frac{1 - e^{-\lambda(k-1)/r}}{n}.$$

Simplify the sum.

On past experience, I know that I will be so flustered that I will just punch in possible integers at random, without noticing which I have already tried. Show that the probability of the police arriving before I get my money is

$$1 - \frac{1}{n - (n-1)e^{-\lambda/r}}.$$

2002 Paper I

Comments

This was originally about getting money from a cash dispenser using a stolen card, but it was decided that STEP questions should not be immoral. Hence the rather more improbable scenario.

The *exponential distribution*, here governing the police arrival time, is often used for failure of equipment (light-bulbs, etc). It has the useful property (not used here) that the reliability for a light bulb (here the probability of the police not coming within a certain time period of duration t) doesn't depend on which time period you choose; i.e. given that the light bulb has survived to time T, the probability of it surviving until time $T + t$ is independent of T (which doesn't seem very suitable for light bulbs). This is referred to as the *memoryless property*.

To simplify the sums, you have to recognise a geometric series where the common ratio of terms is an exponential; not difficult, but easy to miss the first time you see it.

Solution to problem 67

The probability that I get the right number on the kth go is $1/n$. The time taken to key in k integers[33] is $(k-1)/r$. The probability that the police arrive before this is $1 - e^{-\lambda(k-1)/r}$ so the total probability that the police arrive before I get my money is

$$\sum_{k=1}^{n} \frac{1}{n} \times \left(1 - e^{-\lambda(k-1)/r}\right).$$

We have

$$\sum_{k=1}^{n} \frac{1}{n} \times \left(1 - e^{-\lambda(k-1)/r}\right) = 1 - \sum_{k=1}^{n} \frac{e^{-\lambda(k-1)/r}}{n} = 1 - \frac{1 - e^{-\lambda n/r}}{n(1 - e^{-\lambda/r})},$$

since the last part of the sum is a geometric progression with common ratio $e^{-\lambda/r}$.

This time, the probability of getting my money on the kth go is $1/n$ times the probability of not having punched in the correct integer in the preceding $k - 1$ turns:

$$\text{P(money on } k\text{th go)} = \frac{1}{n} \times \left(\frac{n-1}{n}\right)^{k-1}.$$

Thus the probability of the police arriving before I get my money is

$$\sum_{k=1}^{\infty} \frac{1}{n} \times \left(\frac{n-1}{n}\right)^{k-1} \times \left(1 - e^{-\lambda(k-1)/r}\right) = \frac{1}{n} \times \frac{1}{1 - (n-1)/n} - \frac{1}{n} \times \frac{1}{1 - e^{-\lambda/r}(n-1)/n},$$

which is easily seen to be the given answer.

Post-mortem

For each part, you have to calculate the probability of the first success occurring on the kth go. You have to remember to multiply the probability of success on this go by the probability of failure on the preceding $k - 1$ goes — a very typical idea in this sort of probability question. After that, the question is really just algebra. STEP questions on this sort of material nearly always involve significant algebra or calculus.

Car batteries as well as light bulbs often crop up as examples of objects to which exponential reliability can be applied. Suppose your car battery has a guarantee of three years, and your car is completely destroyed after two years. Your insurance company offers to give you one third of the price of the battery. What do you think they would say to your counterclaim for the full price, based on the fact that, given it had lasted two years, the exponential model says that it would have had an expected further three years life in it?

---

[33] Think of fences and posts: k posts here but only $k - 1$ fences.

Problem 68: Breaking plates (✓)

Four students, one of whom is a mathematician, take turns at washing up over a long period of time. The number of plates broken by any student in this time obeys a Poisson distribution, the probability of any given student breaking n plates being $e^{-\lambda}\lambda^n/n!$ for some fixed constant λ, independent of the number of breakages by other students. Given that five plates are broken, find the probability that three or more were broken by the mathematician.

2001 Paper I

Comments

The way this is set up, it is largely a counting exercise (but see postmortem). To start with, you work out the probability of five breakages, then follow that with the probability that the mathematician broke three of more plates. You need to calculate the number of ways that 5 plates can be shared amongst four students, for which you have to consider each *partition* of the number 5 and the number of different ways it can arise.

Solution to problem 68

First we work out the probability of five breakages.

Let P(5, 0, 0, 0) denote the probability that student A breaks 5 plates and students B, C and D break no plates. Then

$$P(5,0,0,0) = \text{(Prob. of breaking 5)} \times \text{(Prob of breaking none)}^3 = \frac{e^{-\lambda}\lambda^5}{5!}\left(e^{-\lambda}\right)^3 = \frac{\lambda^5 e^{-4\lambda}}{5!}.$$

The probability that one student breaks all the plates is $4 \times \lambda^5 e^{-4\lambda}/5!$, the factor 4 because it could be any one of the four students.

Let P(4, 1, 0, 0) denote the probability that student A breaks 4 plates, student B breaks one plate and students C and D break no plate. Then

$$\begin{aligned}P(4,1,0,0) &= \text{(Prob. of breaking 4)} \times \text{(Prob. of breaking one)} \times \text{(Prob of breaking none)}^2 \\ &= \frac{e^{-\lambda}\lambda^4}{4!}\frac{e^{-\lambda}\lambda}{1!}\left(e^{-\lambda}\right)^2 = \frac{\lambda^5 e^{-4\lambda}}{4!}.\end{aligned}$$

The probability that any one student breaks four plates and any other student breaks one plate is therefore $4 \times 3 \times \lambda^5 e^{-4\lambda}/4!$.

Considering P(3, 2, 0, 0), P(3, 1, 1, 0), P(2, 2, 1, 0) and P(2, 1, 1, 1) in turn, together with the number of different ways these probabilities can occur, shows that the probability of five breakages is

$$\lambda^5 e^{-4\lambda}\left(\frac{4}{5!} + \frac{12}{4!\,1!} + \frac{12}{3!\,2!} + \frac{12}{3!\,1!\,1!} + \frac{12}{2!\,2!\,1!} + \frac{4}{2!\,1!\,1!\,1!}\right) = \lambda^5 e^{-4\lambda}\left(\frac{1024}{5!}\right).$$

The probability that the mathematician (student A, say) breaks three or more is P(5, 0, 0, 0)+3P(4, 1, 0, 0)+ 3P(3, 2, 0, 0) + 3P(3, 1, 1, 0), i.e.

$$\lambda^5 e^{-4\lambda}\left(\frac{4}{5!} + \frac{3}{4!\,1!} + \frac{3}{3!\,2!} + \frac{4}{3!\,1!\,1!}\right) = \lambda^5 e^{-4\lambda}\left(\frac{106}{5!}\right).$$

The probability that the mathematician breaks three or more, given that five are broken is therefore $106/1024$.

Post-mortem

If you did the question by the method suggested above, you will probably be wondering about the answer: why is it independent of λ; and why is the denominator 4^5. You will quickly decide that the Poisson distribution was a red herring (though I promise you that it was not an intentional herring), since there is no trace of the distribution in the answer.

The denominator is pretty suggestive. A completely different approach is as follows, It is clear that the probability that the mathematician breaks any given plate is $\frac{1}{4}$. The probability that he or she breaks k plates is therefore binomial:

$$\binom{5}{k}\left(\frac{1}{4}\right)^k\left(\frac{3}{4}\right)^{5-k}$$

and adding the cases $k = 3$, $k = 4$ and $k = 5$ gives rather rapidly

$$\frac{90 + 15 + 1}{1024}.$$

For the method given in the solution, we could substitute P_k for $e^{-\lambda}\lambda^k/k!$ (as the probability of breaking k plates) and the calculation would work just the same, with all the P_ks disappearing; try it.

Problem 69: Lottery (✓✓)

The national lottery of Ruritania is based on the positive integers from 1 to N, where N is very large and fixed. Tickets cost £1 each. For each ticket purchased, the punter (i.e. the purchaser) chooses a number from 1 to N. The winning number is chosen at random, and the jackpot is shared equally amongst those punters who chose the winning number.

A syndicate decides to buy N tickets, choosing every number once to be sure of winning a share of the jackpot. The total number of tickets purchased in this draw is $3.8N$ and the jackpot is £W. Assuming that the non-syndicate punters choose their numbers independently and at random, find the most probable number of winning tickets and show that the expected net loss of the syndicate is approximately

$$N - \frac{5(1-e^{-2.8})}{14} W.$$

2001 Paper II

Comments

This is a binomial distribution problem: the probability that n out of m punters choose the winning ticket is

$$\binom{m}{n} p^n q^{m-n}$$

where here $m = 2.8N$, $p = 1/N$ and $q = 1 - p$. It is clear that an approximation to the binomial distribution is expected (for example, the question uses the word 'approximately' and you have to think about how an approximation might arise); and the presence of the exponential in the given result gives a pretty broad hint that it should be a Poisson distribution — the use of which has to be justified. One can expect the Poisson approximation to work when the number of trials (call it m) is large (e.g. $m > 150$) and when $np \approx npq$, i.e. the mean is roughly equal to the variance (since these are equal for the Poisson distribution) — so $q \approx 1$.

You can't find the most probable number of winning tickets by differentiation (unless you fancy differentiating the factorial x!); instead, you must look at the ratios of consecutive terms and see when these turn from being greater than one to less than one.

Solution to problem 69

Let X be the random variable whose value is the number of winning tickets out of the $2.8N$ tickets purchased by the non-syndicate punters. Then $X \sim \text{Poisson}(2.8)$.

Let $p_j = P(X = j)$. Then

$$p_j = \frac{(2.8)^j e^{-2.8}}{j!} \Rightarrow \frac{p_{j+1}}{p_j} = \frac{2.8}{j+1}.$$

This fraction is greater than 1 if $j < 1.8$, so the most probable number of winning tickets by non-syndicate punters is 2. Overall (including the ticket bought by the syndicate), the most probable number of winning tickets is 3, which is very plausible.

The expected winnings of the syndicate is

$$\left(W p_0 + \frac{1}{2} W p_1 + \frac{1}{3} W p_2 + \cdots \right) = e^{-2.8} W \left(1 + \frac{1}{2} \times \frac{1}{1!}(2.8) + \frac{1}{3} \times \frac{1}{2!}(2.8)^2 + \cdots \right)$$

$$= e^{-2.8} W \frac{e^{2.8} - 1}{2.8},$$

so the expected loss is as given (note that $2.8 = \frac{14}{5}$).

Post-mortem

This is rather interesting. You might perhaps have considered whether it would be worth borrowing money to buy every single lottery combination, in order to win a share of the jackpot. Clearly, the people who run the lottery have to think about this sort of thing.

Suppose the number of tickets sold, excluding the N that we plan to buy, is expected to be kN. Suppose also that a fraction α of the total is paid out in the jackpot. Then, setting $2.8 = k$ and $W = \alpha(k+1)N$, the expected loss formula given in the question becomes

$$N \left(1 - \frac{1 - e^{-k}}{k} \alpha(k+1) \right).$$

If k is very small (take $k = 0$), we lose $N(1 - \alpha)$ (obviously). If k is large, $k = 3$ say, the exponential can be ignored and we lose $N(1 - \alpha(k+1)/k)$. If $k \gg 1$, this becomes $N(1 - \alpha)$ again. In between, there is a value of k that, for each fixed α, gives a minimum loss (which may be a gain if α is close to 1).

Note how informative it is to have k rather than 2.8; the numerical value was chosen in the question to model roughly that lottery system in the UK.

Having gone back to this solution after a break, I am now wondering about the use of the Poisson approximation. Of course, the set-up (large m, small p) begs us to approximate, but did we need to? Certainly not for the first result, since we can just as well look at the ratio of two terms of the Binomial distribution as at two terms of the Poisson distribution. Try it; the result is of course the same.

The second part is more difficult. What we want is the expectation of $1/(n+1)$, and this turns out to be a difficult sum using the Binomial distribution (in fact, it can only be expressed in terms of a hypergeometric function, which would then have to be approximated to get a less obscure answer).

It seems to me that the solution above is therefore a bit unsatisfactory. It would surely have been better to work with the exact distribution until it was necessary to approximate, even though one knows that the approximation is so good that the answers would be the same. We are, after all, mathematicians and not engineers.

Problem 70: Bodies in the fridge (✓✓)

My two friends, who shall remain nameless, but whom I shall refer to as P and Q, both told me this afternoon that there is a body in my fridge. I'm not sure what to make of this, because P tells the truth with a probability of p, while Q (independently) tells the truth with a probability of only q. I haven't looked in the fridge for some time, so if you had asked me this morning, I would have said that there was just as likely to be a body in the fridge as not. Clearly, in view of what my friends have told me, I must revise this estimate. Explain carefully why my new estimate of the probability of there being a body in the fridge should be

$$\frac{pq}{1-p-q+2pq}.$$

I have now been to look in the fridge and there is indeed a body in it; perhaps more than one. It seems to me that only my enemy E_1 or my other enemy E_2 or (with a bit of luck) both E_1 and E_2 could be in my fridge, and this evening I would have judged these three possibilities equally likely. But tonight I asked P and Q separately whether E_1 was in the fridge, and they each said that she was. What should be my new estimate of the probability that both E_1 and E_2 are in my fridge?

Of course, I always tell the truth.

1987 Paper II

Comments

The most difficult part of this problem is unravelling the narrative! The first paragraph says essentially 'what is the probability that there is body in the fridge, given that P and Q both say there is?'. It can therefore be tackled by the usual methods of conditional probability: tree diagrams, for example, or Bayes' theorem. All the other words in the first paragraph are there to tell you about the *a priori* probabilities of the events, without knowledge of which the question above would be meaningless.

In the second paragraph, the situation becomes more complicated, but the method used for the first paragraph will still work.

In case you want to use it, here is the statement of Bayes' theorem, in its simplest form:

$$P(B|A) = \frac{P(B) \times P(A|B)}{P(A)}.$$

Solution to problem 70

This problem can be solved using tree diagrams. A more sophisticated, but not necessarily better, method is to use Bayes' theorem.

Here, we take the events A and B to be

$$A = P \text{ and } Q \text{ both say that there is a body in the fridge}$$
$$B = \text{there is a body in the fridge}$$

From the information given in the question, $P(B) = \tfrac{1}{2}$, so

$$P(B|A) = \frac{\tfrac{1}{2} \times pq}{P(A)}.$$

Now

$$\begin{aligned} P(A) &= P(\text{there is a body}) \times P(P \text{ and } Q \text{ both say there is}) \\ &\quad + P(\text{there is not a body}) \times P(\text{they both say there is}) \\ &= \tfrac{1}{2} \times pq + \tfrac{1}{2} \times (1-p)(1-q) \end{aligned}$$

which gives the required answer.

For the second paragraph, let

$$X = P \text{ and } Q \text{ both say that } E_1 \text{ is in the fridge}$$
$$Y = E_1 \text{ and } E_2 \text{ are in the fridge}$$

There are three possibilities (since we know that there is at least one body in the fridge): only E_1 is in the fridge; only E_2 is in the fridge; and both E_1 and E_2 are in the fridge. These are given as equally likely, so the *a priori* probabilities are each $\tfrac{1}{3}$.

We want $P(Y|X)$, which by Bayes' theorem is

$$\frac{P(Y) \times P(X|Y)}{P(X)} = \frac{\tfrac{1}{3} \times pq}{P(X)}.$$

Now

$$\begin{aligned} P(X) &= P(\text{only } E_1 \text{ is in the fridge}) \times P(P \text{ and } Q \text{ told the truth}) \\ &\quad + P(\text{both } E_1 \text{ and } E_2 \text{ are in the fridge}) \times P(P \text{ and } Q \text{ both told the truth}) \\ &\quad + P(\text{only } E_2 \text{ is in the fridge}) \times P(P \text{ and } Q \text{ both lied}) \\ &= \tfrac{1}{3} \times pq + \tfrac{1}{3} \times pq + \tfrac{1}{3} \times (1-p)(1-q) \end{aligned}$$

so the answer is $\dfrac{pq}{1 - p - q + 3pq}.$

Post-mortem

Note how much more difficult it is when the answer is not given; when the question was originally set, most candidates arrived at the given answer to the first part but were not sufficiently confident to extend their method to the second paragraph: they received $\tfrac{8}{20}$ for their efforts.

The last line of the question is not entirely frivolous; if I may have lied about what my friends answered when I asked them if there is a body in the fridge, the problem becomes difficult. However, my claim to be truthful is vacuous (it tells you nothing) because I may be lying. Contrast with the statement 'I am lying', which is inconsistent.

Problem 71: Choosing keys (✓)

> I have k different keys on my key ring. When I come home at night I try one key after another until I find the key that fits my front door. What is the probability that I find the correct key, for the first time, on the nth attempt in each of the following three cases?
>
> (i) At each attempt, I choose a key that I have not tried before, each choice being equally likely.
>
> (ii) At each attempt, I choose a key from all my keys, each of the k choices being equally likely.
>
> (iii) At the first attempt, I choose from all my keys, each of the k choices being equally likely. At each subsequent attempt, I choose from the keys that I did not try at the previous attempt, each of the $k-1$ choices being equally likely.

2000 Paper II

Comments

This is very easy, and would really be far too easy if the answers were given.

You should set out your argument clearly and concisely, because if you come up with the wrong answer and an inadequate explanation you will not get many marks. Even if you write down the correct answer you may not get the marks if your explanation is inadequate.

You should of course run the usual checks on your answers. Do they lie in the range $0 \leqslant p \leqslant 1$? If you sum over all outcomes, do you get 1? (The latter is a good and not difficult exercise; I insist that you try it.)

You will probably be struck by the simplicity of the answer to part (i), after simplification. Is there an easy way of thinking about it (or, if you did it the easy way, is there a hard way)?

Solution to problem 71

(i)

P(finding correct key, for the first time, on nth attempt)

= P(fail first, fail second, ... , fail $(n-1)$th, succeed nth)

$$= \begin{cases} \dfrac{1}{k} & \text{for } n = 1; \\ \dfrac{k-1}{k} \times \dfrac{k-2}{k-1} \times \cdots \times \dfrac{k-(n-1)}{k-(n-2)} \times \dfrac{1}{k-(n-1)} = \dfrac{1}{k} & \text{for } 2 \leqslant n \leqslant k; \\ 0 & \text{for } n > k. \end{cases}$$

(ii)

P(find correct key, for the first time, on nth attempt)

= P(fail first, fail second, ... , succeed nth) = $\left(\dfrac{k-1}{k}\right)^{n-1} \dfrac{1}{k}$.

(iii)

P(find correct key, for the first time, on nth attempt)

= P(fail first, fail second, ... , succeed nth)

$$= \begin{cases} \dfrac{1}{k} & \text{for } n = 1; \\ \left(\dfrac{k-1}{k}\right)\left(\dfrac{k-2}{k-1}\right)^{n-2} \dfrac{1}{k-1} & \text{for } n \geqslant 2. \end{cases}$$

Post-mortem

The above solution for part (i) is the 'hard way'. The easy way is to consider the keys all laid out in a row, instead of being picked sequentially. Exactly one of the keys is the correct one, which is equally likely to be any of the keys in the row, so has a probability of $1/k$ of being in any given position. Clearly, this corresponds exactly to picking them one by one.

I gave the hard solution above, because I think most people will be drawn into doing it this way by the phrasing of the question: keys are picked one by one and tried before going on to the next.

Did you check that your probabilities sum to 1? It is obvious for part (i), but you have to sum geometric progressions for parts (ii) and (iii).

Problem 72: Commuting by train (✓✓)

Tabulated values of $\Phi(\cdot)$, the cumulative distribution function of a standard normal variable, should not be used in this question.

Henry the commuter lives in Cambridge and his working day starts at his office in London at 0900. He catches the 0715 train to King's Cross with probability p, or the 0720 to Liverpool Street with probability $1 - p$. Measured in minutes, journey times for the first train are $N(55, 25)$ and for the second are $N(65, 16)$. Journey times from King's Cross and Liverpool Street to his office are $N(30, 144)$ and $N(25, 9)$, respectively. Show that Henry is more likely to be late for work if he catches the first train.

Henry makes M journeys, where M is large. Writing A for $1 - \Phi(\frac{20}{13})$ and B for $1 - \Phi(2)$, find, in terms of A, B, M and p, the expected number, L, of times that Henry will be late and show that, for all possible values of p,

$$BM \leqslant L \leqslant AM.$$

Henry noted that in $\frac{3}{5}$ of the occasions when he was late, he had caught the King's Cross train. Obtain an estimate of p in terms of A and B.

Note: A random variable is said to be $N(\mu, \sigma^2)$ if it has a normal distribution with mean μ and variance σ^2.

2000 Paper II

Comments

This is impossible unless you know the following result:

If X_1 and X_2 are independent and normally distributed according to $X_1 \sim N(\mu_1, \sigma_1^2)$ and $X_2 \sim N(\mu_2, \sigma_2^2)$, then $X_1 + X_2$ is also normally distributed and $X_1 + X_2 \sim N(\mu_1 + \mu_2, \sigma_1^2 + \sigma_2^2)$.

Even if you didn't know this result, you do now and it should not be difficult to complete the first parts question.

For the last part, you need to know something about conditional probability, namely that

$$P(A|B) = \frac{P(B \cap A)}{P(B)}$$

which makes sense intuitively and can easily be understood in terms of Venn diagrams. The denominator is essentially a normalising constant. The formula may be taken as the definition of conditional probability on the left hand side.

Solution to problem 72

Let T_1 be the random variable representing the total journey time via Kings Cross, so that

$$T_1 \sim N(55 + 30, 25 + 144) = N(85, 169),$$

using the result mentioned on the previous page, and and let T_2 be the random variable representing the total journey time via Liverpool Street, so that

$$T_2 \sim N(65 + 25, 16 + 9) = N(90, 25).$$

Then the probabilities of being late are, respectively, $P(T_1 > 105)$ and $P(T_2 > 100)$, i.e. $1 - \Phi(\frac{20}{13})$ and $1 - \Phi(2)$. Note that $\Phi(\frac{20}{13}) < \Phi(2)$.

We have

$$L = [pA + (1-p)B]M = [B + (A-B)p]M.$$

L increases as p increases, since $A > B$, hence the given inequalities corresponding to $p = 0$ and $p = 1$.

We have

$$P(\text{Kings Cross given late}) = \frac{P(\text{Late and Kings Cross})}{P(\text{Late})}$$

An estimate for p (all it $\tilde{p}$) is therefore given by

$$\frac{3}{5} = \frac{A\tilde{p}}{A\tilde{p} + B(1-\tilde{p})},$$

so $\tilde{p} = \dfrac{3B}{2A + 3B}$.

Post-mortem

The result mentioned in the comment on the previous page is just the sort of thing you should try to prove yourself rather than take on trust. Unfortunately, such results in probability tend to be pretty hard to prove. You can also prove it fairly easily using generating functions (which are not in the syllabus for STEP I and II). You can also prove it from first principles. That would be difficult for you because of the awkward integrals involved, but not impossible.

Problem 73: Collecting voles (✓)

A group of biologists attempts to estimate the magnitude, N, of an island population of voles (*Microtus agrestis*). Accordingly, the biologists capture a random sample of 200 voles, mark them and release them. A second random sample of 200 voles is then taken of which 11 are found to be marked. Show that the probability, p_N, of this occurrence is given by

$$p_N = k\frac{((N-200)!)^2}{N!(N-389)!},$$

where k is independent of N.

The biologists then estimate N by calculating the value of N for which p_N is a maximum. Find this estimate.

All unmarked voles in the second sample are marked and then the entire sample is released. Subsequently a third random sample of 200 voles is taken. Using your estimate for N, write down the probability that this sample contains exactly j marked voles, leaving your answer in terms of binomial coefficients.

Deduce that

$$\sum_{j=0}^{200} \binom{389}{j}\binom{3247}{200-j} = \binom{3636}{200}.$$

2000 Paper II

Comments

This is really just an exercise in combinations. (Recall that a permutation is a reordering of a set of objects, and a combination is a selection of a subset from a set.) You assume that you are equally likely to choose any given subset of the same size, so that the probability of a set of specific composition is the number of ways of choosing a set of that composition divided by the total number of ways of choosing any set of the same size. Of course, you are assuming that the voles are indistinguishable, except for the marks made by the biologists.

Maximising a discrete (not a continuous) function of N came up on one of the previous questions: you have to compare adjacent terms.

The numbers look rather bad, though they turn out OK. My instinct would be to do it algebraically first: replace 200 by a and 11 by b, then substitute back at the end. I am sure it will lead to a better understanding of what is going on.

Solution to problem 73

For the second sample, 200 out of N voles are already marked, so p_N is just the number of ways of choosing 11 from 200 and 189 from $N - 200$ divided by the number of ways of choosing 200 from N:

$$p_N = \frac{\binom{200}{11}\binom{N-200}{189}}{\binom{N}{200}} = \frac{\frac{200!}{11!\,189!}\,\frac{(N-200)!}{189!\,(N-389)!}}{\frac{N!}{200!\,(N-200)!}} \, ,$$

so

$$k = \frac{(200!)^2}{11!\,(189!)^2} \, .$$

At the maximum value, $p_N \approx p_{N-1}$, i.e.

$$\frac{(N-200)^2}{N(N-389)} \approx 1 \, ,$$

which gives $N \approx 200^2/11 \approx 3636$ (just divide 40000 by 11).

At the third sample, there are 389 marked voles and an estimated $3636 - 389 = 3247$ unmarked voles. Hence

$$P(\text{exactly } j \text{ marked voles}) = \frac{\binom{389}{j}\binom{3247}{200-j}}{\binom{3636}{200}} \, .$$

The final part follows immediately, using

$$\sum_{j=0}^{200} P(\text{exactly } j \text{ marked voles}) = 1 \, .$$

Post-mortem

Adding the Latin name of the species was a nice touch, I thought (not my idea); it adds an air of verisimilitude to the problem.

I suppose that this might be the basis of a method of estimating the population of voles — rather clever really. I don't know how well it works in practice though. The assumption mentioned earlier, that picking one set of voles of size 200 is just as likely as picking any other set, surely relies on perfect mixing of the marked voles, which would be rather difficult to achieve (especially as female voles can be highly territorial).

You might be asking yourself why it was OK to find the maximum by setting $p_N \approx p_{N-1}$. This is a standard method, but of course it only works if the distribution is one-humped, like a normal distribution. An alternative approach would have been to approximate the distributions using Stirling's approximation, which at its most basic is

$$\ln N! \approx N \ln N - N \, .$$

This gives exactly the same equation as the $p_N \approx p_{N-1}$ method, and shows that the distribution is indeed one-humped.

Problem 74: Breaking a stick (✓✓)

> A stick is broken at a point, chosen at random, along its length. Find the probability that the ratio, R, of the length of the shorter piece to the length of the longer piece is less than r, where r is a given positive number.
>
> Find the probability density function for R, and calculate the mean and variance of R.

1999 Paper II

Comments

Continuous probability distributions often seem harder than discrete distributions, for no good reason: the concepts are the same, and in fact integrals are normally easier than sums.

Here, part of the difficulty is that you have to set up the problem yourself. You have a random variable with a known distribution (corresponding to the point at which the stick is broken), but you are interested in another random variable (the ratio of lengths) derived from the first. As always, it is best to work with the cumulative distribution functions rather than the probability density functions when deriving the distribution of the second random variable.

Solution to problem 74

Let the length of the stick be 2ℓ and let X be the length of the shorter piece of stick, so that $X \leqslant \ell$. The random variable X is uniformly distributed on the interval $0 \leqslant x \leqslant \ell$, so

$$P(0 \leqslant X \leqslant x) = \frac{x}{\ell}.$$

Now $R = \dfrac{X}{2\ell - X}$, by definition, so

$$X = \frac{2\ell R}{1 + R}.$$

The cumulative distribution function for R is given by

$$P(R \leqslant r) = P\left(\frac{X}{2\ell - X} \leqslant r\right) = P\left(X \leqslant \frac{2\ell r}{1+r}\right) = \frac{2\ell r}{(1+r)\ell} = \frac{2r}{1+r}.$$

Let's check that this satisfies the conditions for a cumulative distribution function: it should increase from 0 to 1 as r goes from its smallest value, which is 0 to its greatest value, which is 1. And it does, so that's OK.

The probability density function is the derivative of the cumulative distribution function:

$$\frac{d}{dr}\left(\frac{2r}{1+r}\right) = \frac{2}{(r+1)^2}.$$

Integrating to find $E(R)$ and $E(R^2)$ gives

$$E(R) = \int_0^1 \frac{2r}{(1+r)^2}\,dr = 2\ln 2 - 1; \quad E(R^2) = \int_0^1 \frac{2r^2}{(1+r)^2}\,dr = 3 - 4\ln 2,$$

and $\text{Var}\,(R) = (3 - 4\ln 2) - (2\ln 2 - 1)^2 = 2 - 4(\ln 2)^2$.

Post-mortem

Note the check we made to verify that the function we found for the cumulative distribution could actually be a cumulative distribution function, which is equivalent to checking that the probability density function integrates to 1.

We should also check that the final answers make sense. Putting numbers into a calculator gives $2\ln 2 - 1 = 0.23$ for the expected value, which seems reasonable; at least it is less than 1. If you sketch the density function, you see that this could easily be its average value (do it!).

For the variance, I got 0.08, which means that the ratio is likely to be in the plausible range 0.15 to 0.31.

There are a variety of other stick-breaking problems, including a rather pleasing one about breaking the stick in two places and finding the probability that the pieces form a triangle. In this case, it matters how you break the stick: you could break it once and then break one of the pieces; or you could choose two points on the stick randomly at which to break it. For an interesting discussion, see the excellent cut-the-knot web site: http://www.cut-the-knot.org/Curriculum/Probability/TriProbability.shtml

Problem 75: Random quadratics (✓✓✓)

The random variable B is normally distributed with mean zero and unit variance. Find the probability that the quadratic equation

$$X^2 + 2BX + 1 = 0$$

has real roots.

Given that the two roots X_1 and X_2 are real, find, giving your answers to three significant figures:

(i) the probability that both X_1 and X_2 are greater than $\frac{1}{5}$;

(ii) the expected value of $|X_1 + X_2|$.

1988 Paper II

Comments

It is quite difficult to find statistics questions at this level that are not too difficult and are also not simple applications of standard methods. For example, χ^2 tests are not really suitable, because the theory is sophisticated while the applications are usually rather straightforward. Most questions in the probability/statistics area tend therefore to concentrate on probability, and many of these have a bit of pure mathematics thrown in.

Here, the random variable is the coefficient of a quadratic equation, which is rather pleasing. But you have to handle the inequalities carefully. The difficulty is increased by the conditional element: for parts (i) and (ii) you are only interested in the case of real roots.

If you don't have statistical tables handy, don't bother to find some: just leave the answers in terms of the probability, $\Phi(z)$, that a standard ($\mu = 0$, $\sigma = 1$) normally distributed random variable Z satisfies $Z \leqslant z$.

Solution to problem 75

The solution of the quadratic is
$$X = -B \pm \sqrt{B^2 - 1}$$

which has real roots if $|B| \geqslant 1$. Let $\Phi(z)$ be the probability that a standard normally distributed variable Z satisfies $Z \leqslant z$. Then the probability that $|B| \geqslant 1$ is (taking the two tails of the normal distribution)
$$\Phi(-1) + (1 - \Phi(1)) = 2 - 2\,\Phi(1) = 0.3174.$$

(i) We need the smaller root to be greater than $\tfrac{1}{5}$. The smaller root is $-B - \sqrt{B^2 - 1}$. Now provided $\sqrt{B^2 - 1}$ is real, we have

$$-B - \sqrt{B^2 - 1} > \tfrac{1}{5} \quad \Leftrightarrow \quad B + \tfrac{1}{5} < -\sqrt{B^2 - 1}$$
$$\Leftrightarrow \quad (B + \tfrac{1}{5})^2 > B^2 - 1 \quad \text{and} \quad (B + \tfrac{1}{5}) < 0$$
$$\Leftrightarrow \quad \tfrac{2}{5}B + \tfrac{1}{25} > -1 \quad \text{and} \quad B < -\tfrac{1}{5}$$
$$\Leftrightarrow \quad -\tfrac{1}{5} > B > -\tfrac{13}{5}.$$

However, if $B < -\tfrac{1}{5}$, then the condition that $\sqrt{B^2 - 1}$ is real, i.e. $|B| \geqslant 1$, implies the stronger condition $B \leqslant -1$. The condition that both roots are real *and* greater than $\tfrac{1}{5}$ is therefore

$$-\tfrac{13}{5} < B \leqslant -1$$

and the probability that both roots are real and greater than $\tfrac{1}{5}$ is

$$\Phi(-1) - \Phi(-2.6) = \Phi(2.6) - \Phi(1) = 0.9953 - 0.8413 = 0.1540.$$

The conditional probability that both roots are greater than $\tfrac{1}{5}$ given that they are real is

$$P\bigl(\text{both roots} > \tfrac{1}{5} \mid \text{both real}\bigr) = \frac{P\bigl(\text{both roots} > \tfrac{1}{5} \text{ and both roots real}\bigr)}{P\bigl(\text{both roots real}\bigr)}$$

$$= \frac{0.1540}{0.3174} = 0.485.$$

(ii) The sum of the roots is $-2B$, so we want the expectation of $|2B|$ given that $|B| \geqslant 1$, which is

$$\frac{\dfrac{1}{\sqrt{2\pi}}\displaystyle\int_{-\infty}^{-1}(-2x)e^{-\tfrac{1}{2}x^2}\,dx + \dfrac{1}{\sqrt{2\pi}}\displaystyle\int_{1}^{\infty}2xe^{-\tfrac{1}{2}x^2}\,dx}{\dfrac{1}{\sqrt{2\pi}}\displaystyle\int_{-\infty}^{-1}e^{-\tfrac{1}{2}x^2}\,dx + \dfrac{1}{\sqrt{2\pi}}\displaystyle\int_{1}^{\infty}e^{-\tfrac{1}{2}x^2}\,dx} = \frac{2 \times \dfrac{1}{\sqrt{2\pi}} \times 2e^{-\tfrac{1}{2}}}{2(1 - \Phi(1))} = 3.05.$$

Syllabus

The mathematical requirements for this book are based on the syllabus for STEP Mathematics I and II. The syllabus listed below serves as a rough guide. Some of the questions in the book require no knowledge of the syllabus; and some cover material that is not included in the syllabus, but is introduced in the question itself. You can find more information about the STEP examinations from the web site http://www.stepmathematics.org.uk.

PURE MATHEMATICS

Specification	Notes
General	
Mathematical vocabulary and notation	including: equivalent to; necessary and sufficient; if and only if; $\Rightarrow$; $\Leftrightarrow$; $\equiv$.
Methods of proof	including proof by contradiction and disproof by counterexample; proof by induction.
Algebra	
Indices and surds	including rationalising denominators.
Quadratics	including proving positivity by completing a square.
The expansion for $(a+b)^n$	including knowledge of the general term; notation: $\binom{n}{r} = \dfrac{n!}{r!\,(n-r)!}$.
Algebraic operations on polynomials and rational functions	including factorisation, the factor theorem, the remainder theorem; including understanding that, for example, if $$x^3 + bx^2 + cx + d \equiv (x-\alpha)(x-\beta)(x-\gamma),$$ then $d = -\alpha\beta\gamma$.
Partial fractions	including denominators with a repeated or quadratic factor.
Sequences and series	including use of, for example, $a_{n+1} = \mathrm{f}(a_n)$ or $a_{n+1} = \mathrm{f}(a_n, a_{n-1})$; including understanding of the terms convergent, divergent and periodic in simple cases; including use of $\sum_{k=1}^{n} k$ to obtain related sums.
The binomial series for $(1+x)^k$, where k is a rational number	including understanding of the condition $|x| < 1$.
Arithmetic and geometric series	including sums to infinity and conditions for convergence, where appropriate.

170 Advanced Problems in Mathematics

Inequalities	including solution of, eg, $\frac{1}{a-x} > \frac{x}{x-b}$; including simple inequalities involving the modulus function; including the solution of simultaneous inequalities by graphical means.

Functions

Domain, range, composition, inverse	including use of functional notation such as $y = \mathrm{f}(ax+b)$, $x = \mathrm{f}^{-1}(y)$ and $z = \mathrm{f}(\mathrm{g}(x))$.
Increasing and decreasing functions	the precise definition of these terms will be included in the question.
Exponentials and logarithms	including $x = a^y \Leftrightarrow y = \log_a x$, $x = \mathrm{e}^y \Leftrightarrow y = \ln x$; including the exponential series
The effect of simple transformations	such as $y = a\mathrm{f}(bx+c) + d$.
The modulus function.	
Location of roots of $\mathrm{f}(x) = 0$ by considering changes of sign of $\mathrm{f}(x)$	
Approximate solution of equations using simple iterative methods	

Curve sketching

General curve sketching	including use of symmetry, transformations, behaviour as $x \to \pm\infty$, points or regions where the function is undefined, turning points, asymptotes parallel to the axes.

Trigonometry

Radian measure, arc length of a circle, area of a segment	
Trigonometric functions	including knowledge of standard values, such as $\tan(\frac{1}{4}\pi)$, $\sin 30°$; including identities such as $\sec^2\phi - \tan^2\phi = 1$; including application to geometric problems in two and three dimensions.
Double angle formulae	including their use in calculating, eg, $\tan(\frac{1}{8}\pi)$.
Formulae for $\sin(A \pm B)$ and $\cos(A \pm B)$	including their use in solving equations such as $a\cos\theta + b\sin\theta = c$.
Inverse trigonometric functions	definitions including domains and ranges; notation: either $\arctan\theta$ or $\tan^{-1}\theta$, etc.

Coordinate geometry

Straight lines in two-dimensions	including the equation of a line through two given points, or through a given point and parallel to a given line or through a given point and perpendicular to a given line; including finding a point which divides a segment in a given ratio.
Circles	using the general form $(x-a)^2 + (y-b)^2 = R^2$; including points of intersection of circles and lines.

Cartesian and parametric equations of curves
and conversion between the two forms.

Calculus

Interpretation of a derivative as a limit and as a rate of change	including knowledge of both notations $f'(x)$ and $\frac{dy}{dx}$.
Differentiation of standard functions	including algebraic expressions, trigonometric and inverse trigonometric functions, exponential and log functions.
Differentiation of composite functions, products and quotients and functions defined implicitly.	
Higher derivatives	including knowledge of both notations $f''(x)$ and $\frac{d^2y}{dx^2}$; including knowledge of the notation $\frac{d^ny}{dx^n}$.
Applications of differentiation to gradients, tangents and normals, stationary points, increasing and decreasing functions	including finding maxima and minima which are not stationary points; including classification of stationary points using the second derivative.
Integration as reverse of differentiation	
Integral as area under a curve	including area between two curves; including approximation of integral by the rectangle and trapezium rules.
Volume within a surface of revolution	rotation about either x or y axes.
Knowledge and use of standard integrals	including the forms $\int f'(g(x))g'(x)\,dx$ and $\int f'(x)/f(x)\,dx$; including transformation of an integrand into standard (or some given) form; including use of partial fractions; not including knowledge of integrals involving inverse trigonometric functions.
Definite integrals	including calculation, without justification, of simple improper integrals such as $\int_0^\infty e^{-x}\,dx$ and $\int_0^1 x^{-\frac{1}{2}}\,dx$ (if required, information such as the behaviour of xe^{-x} as $x \to \infty$ or of $x\ln x$ as $x \to 0$ will be given).
Integration by parts and by substitution	including understanding their relationship with differentiation of product and of a composite function; including application to (e.g.) $\int \ln x\,dx$.
Formulation and solution of differential equations	formulation of first order equations; solution in the case of a separable equation or by some other method given in the question.

Vectors

Vectors in two and three dimensions	including use of column vector and **i**, **j**, **k** notation.
Magnitude of a vector	including the idea of a unit vector.
Vector addition and multiplication by scalars	including geometrical interpretations.
Position vectors	including application to geometrical problems.

The distance between two points	
Vector equations of lines	including the finding the intersection of two lines; understanding the notion of skew lines (knowledge of shortest distance between skew lines is not required).
The scalar product	including its use for calculating the angle between two vectors.

MECHANICS

Questions on mechanics may involve any of the material in the Pure Mathematics syllabus.

Specification	Notes
Force as a vector	including resultant of several forces acting at a point and the triangle or polygon of forces; including equilibrium of a particle; forces include weight, reaction, tension and friction.
Centre of mass	including obtaining the centre of mass of a system of particles, of a simple uniform rigid body (possibly composite) and, in simple cases, of non-uniform body by integration.
Equilibrium of a rigid body or several rigid bodies in contact	including use of moment of a force; for example, a ladder leaning against a wall or on a movable cylinder; including investigation of whether equilibrium is broken by sliding, toppling or rolling; including use of Newton's third law; excluding questions involving frameworks.
Kinematics of a particle in a plane	including the case when velocity or acceleration depends on time (but excluding knowledge of acceleration in the form $v\frac{dv}{dx}$); questions may involve the distance between two moving particles, but detailed knowledge of relative velocity is not required.
Energy (kinetic and potential), work and power	including application of the principle of conservation of energy.
Collisions of particles	including conservation of momentum, conservation of energy (when appropriate); coefficient of restitution, Newton's experimental law; including simple cases of oblique impact (on a plane, for example); including knowledge of the terms *perfectly elastic* ($e = 1$) and *inelastic* ($e = 0$); questions involving successive impacts may be set.
Newton's first and second laws of motion	including motion of a particle in two and three dimensions and motion of connected particles, such as trains, or particles connected by means of pulleys.

Motion of a projectile under gravity	including manipulation of the equation $$y = x \tan \alpha - \frac{gx^2}{2V^2 \cos^2 \alpha},$$ viewed, possibly, as a quadratic in $\tan \alpha$; not including projectiles on inclined planes.

PROBABILITY AND STATISTICS

The emphasis is towards probability and formal proofs, and away from data analysis and use of standard statistical tests. Questions may involve use of any of the material in the Pure Mathematics syllabus.

Specification	Notes
Probability	
Permutations, combinations and arrangements	including sampling with and without replacement.
Exclusive and complementary events	including understanding of $P(A \cup B) = P(A) + P(B) - P(A \cap B)$, though not necessarily in this form.
Conditional probability	informal applications, such as tree diagrams.
Distributions	
Discrete and continuous probability density functions and cumulative distribution functions	including calculation of mean, variance, median, mode and expectations by explicit summation or integration for a given (possibly unfamiliar) distribution (eg exponential or geometric or something similarly straightforward); notation: $f(x) = F'(x)$.
Binomial distribution	including explicit calculation of mean.
Poisson distribution	including explicit calculation of mean; including use as approximation to binomial distribution where appropriate.
Normal distribution	including conversion to the standard normal distribution by translation and scaling; including use as approximation to the binomial or Poisson distributions where appropriate; notation: $X \sim N(\mu, \sigma^2)$.
Hypothesis testing	
Basic concepts in the case of a simple null hypothesis and simple or compound alternative	including knowledge of the terminology *null hypothesis* and *alternative hypothesis*, *one* and *two tailed tests*.

This book need not end here...

At Open Book Publishers, we are changing the nature of the traditional academic book. The title you have just read will not be left on a library shelf, but will be accessed online by hundreds of readers each month across the globe. We make all our books free to read online so that students, researchers and members of the public who cant afford a printed edition can still have access to the same ideas as you.

Our digital publishing model also allows us to produce online supplementary material, including extra chapters, reviews, links and other digital resources. Find Advanced Problems in Mathematics on our website to access its online extras. Please check this page regularly for ongoing updates, and join the conversation by leaving your own comments:

http://www.openbookpublishers.com/isbn/9781783741427

If you enjoyed this book, and feel that research like this should be available to all readers, regardless of their income, please think about donating to us. Our company is run entirely by academics, and our publishing decisions are based on intellectual merit and public value rather than on commercial viability. We do not operate for profit and all donations, as with all other revenue we generate, will be used to finance new Open Access publications.

For further information about what we do, how to donate to OBP, additional digital material related to our titles or to order our books, please visit our website: http://www.openbookpublishers.com

You may also be interested in...

http://www.openbookpublishers.com/product/340

Teaching Mathematics is nothing less than a mathematical manifesto. Arising in response to a limited National Curriculum, and engaged with secondary schooling for those aged 11-14 (Key Stage 3) in particular, this handbook for teachers will help them broaden and enrich their students' mathematical education. It avoids specifying how to teach, and focuses instead on the central principles and concepts that need to be borne in mind by all teachers and textbook authors but which are little appreciated in the UK at present.

This study is aimed at anyone who would like to think more deeply about the discipline of 'elementary mathematics', in England and Wales and anywhere else. By analysing and supplementing the current curriculum, *Teaching Mathematics* provides food for thought for all those involved in school mathematics, whether as aspiring teachers or as experienced professionals. It challenges us all to reflect upon what it is that makes secondary school mathematics educationally, culturally, and socially important.

Tony Gardiner, former Reader in Mathematics and Mathematics Education at the University of Birmingham, was responsible for the foundation of the United Kingdom Mathematics Trust in 1996, one of the UK's largest mathematics enrichment programs. In 1997 Gardiner served as President of the Mathematical Association, and in 2011 was elected Education Secretary of the London Mathematical Society.

Lightning Source UK Ltd.
Milton Keynes UK
UKHW051205180919
349970UK00009B/28/P

9 781783 741427